Programmable Logic Controller (PLC) Tutorial, Siemens Simatic S7-200

Circuits and programs for Siemens Simatic S7-200 programmable controllers

for Electrical Engineers and Technicians

Stephen P. Tubbs, P.E.
*formerly of the
Pennsylvania State University,
currently an
industrial consultant*

Copyright 2007 by Stephen P. Tubbs
1344 Firwood Dr.
Pittsburgh, PA 15243

All rights reserved. No part of this book may be reproduced, in any form or by any means, without permission in writing from the publisher.

NOTICE TO THE READER

The author does not warrant or guarantee any of the products, equipment, or programs described herein or accept liability for any damages resulting from their use.

The reader is warned that electricity and the construction of electrical equipment are dangerous. It is the responsibility of the reader to use common sense and safe electrical and mechanical practices.

S7-200, S7-300, S7-400, Step 7, Step 7-Micro/WIN, and Simatic are trademarks of Siemens AG.

GE Fanuc, Nano and VersaMax are trademarks of GE Fanuc Automation.

MicroLogix, PLC, and SLC 500 are trademarks of Allen-Bradley, a division of Rockwell Automation.

Modicon is a trademark of Schneider Electric.

MSWindows is a trademark of Microsoft.

Printed in the United States of America

ISBN   978-0-9659446-8-7

CONTENTS

	PAGE
INTRODUCTION	vii
1.0 SIEMENS FAMILIES OF PLCS	1
2.0 S7-200 FAMILY HARDWARE AND SOFTWARE	3
2.1 S7-200 BASE UNITS	3
2.2 S7-200 OPTIONAL EXPANSION MODULES	4
2.3 S7-200 PROGRAMMING SOFTWARE, "STEP 7-MICRO/WIN"	4
3.0 S7-200 COMMUNICATIONS	7
3.1 EIA-485 PORT	7
3.2 INDUSTRIAL ETHERNET	8
3.3 INTERNET	9
3.4 PROFIBUS DP	9
3.5 AS-INTERFACE	9
3.6 S7-200 PC ACCESS	10
4.0 S7-200 CPU 222 "TEACHING SETUP"	11
4.1 TRAINING PROMOTION STARTER KIT	11

4.2 ALTERNATIVE TO THE TRAINING PROMOTION STARTER KIT   15

5.0 RELAY LADDER CONNECTION AND POWER DIAGRAMS   21

6.0 MAKING "STEP 7-MICRO/WIN" WORK   25

6.1 INSTALLING "STEP 7-MICRO/WIN"   25

6.2 CONNECTING THE CPU 222 "TEACHING SETUP"   25

6.3 VERIFYING AND SETTING "STEP 7-MICRO/WIN" COMMUNICATIONS PARAMETERS   25

6.4 PERSONAL COMPUTER INFORMATION THAT YOU MAY NEED   26

6.5 USING THE S7-200 "RUN/TERM/STOP" SWITCH   27

7.0 EXAMPLE CPU 222 LADDER AND POWER DIAGRAMS   28

7.1 CPU 222 LADDER AND POWER DIAGRAMS FOR THREE-PHASE INDUCTION MOTOR FORWARD AND REVERSE CONTROL   29

7.2 CPU 222 LADDER AND POWER DIAGRAMS FOR TIMED SEQUENTIAL STARTING OF TWO THREE-PHASE INDUCTION MOTORS   39

7.3 CPU 222 LADDER AND POWER DIAGRAMS FOR A COUNTER-CONTROLLED THREE-PHASE INDUCTION MOTOR POWERED BOTTLE PUSHER   44

7.4 CPU 222 LADDER AND POWER DIAGRAMS FOR A COUNTER-CONTROLLED THREE-PHASE INDUCTION MOTOR WITH A SETABLE COUNTER   48

7.5 CPU 222 LADDER AND POWER DIAGRAMS FOR ANOTHER COUNTER-CONTROLLED THREE-PHASE INDUCTION MOTOR   53

7.6 CPU 222 LADDER AND POWER DIAGRAMS FOR TIMED SEQUENTIAL STARTING OF TWO THREE-PHASE INDUCTION MOTORS WITH A PROGRAM CONTROL STOP   58

7.7 CPU 222 LADDER AND POWER DIAGRAMS FOR A RETENTIVE
TIMER CONTROLLED MACHINE   63

7.8 CPU 222 LADDER AND POWER DIAGRAMS FOR A SYSTEM TO
DETERMINE IF A BOTTLE COUNT RATE IS TOO LOW
OR TOO HIGH   67

7.9 CPU 222 LADDER AND POWER DIAGRAMS FOR A SYSTEM
USING BOTTLE COUNT RATE TO SELECT PRESET PUMP
MOTOR INVERTER SPEEDS   71

7.10 CPU 222 LADDER AND POWER DIAGRAMS FOR A CONVEYOR
BELT PART PLACER   76

7.11 CPU 222 LADDER AND POWER DIAGRAMS FOR A CONVEYOR
BELT SPEED CONTROLLER   81

7.12 CPU 222 LADDER AND POWER DIAGRAMS FOR A SYSTEM
THAT USES TABLE STORAGE AND FIFO TO SELECT SPRAY
PAINT COLORS   88

7.13 CPU 222 LADDER DIAGRAM SUBROUTINES AND JUMP
STATEMENT   94

7.14 CPU 222 LADDER AND POWER DIAGRAMS FOR A SYSTEM
THAT USES HIGH-SPEED COUNTING TO MEASURE AND
CONTROL MOTOR SPEED   100

8.0 OTHER SOURCES OF INFORMATION   107

8.1 USEFUL REFERENCE BOOKS   107

8.2 SIEMENS CONTACTS   108

8.3 ELECTRICAL DISTRIBUTORS THAT SELL SIEMENS PLCS   108

9.0 APPENDIX   109

9.1 RELAY LADDER AND POWER DIAGRAM SYMBOLS   109

9.2 "STEP 7-MICRO/WIN" LADDER DIAGRAM PROGRAM MEMORY
    AREAS   111

9.3 "STEP 7-MICRO/WIN" LADDER DIAGRAM PROGRAM
    INSTRUCTION SYMBOLS   112

9.4 PROGRAMMABLE CONTROLLER GLOSSARY   123

9.5 TECHNICAL SPECIFICATIONS COMMON TO TODAY'S S7-200S
    (CPUS 221, 222, 224, 224XP, & 226)   129

9.6 TECHNICAL SPECIFICATIONS SPECIFIC TO THE CPU 222   130

9.7 EXPANSION MODULES AVAILABLE FOR CPUS 222, 224,
    224XP, & 226   131

INTRODUCTION

The purpose of this book is to teach and demonstrate the basics of the Siemens AG Automation and Drives Industrial Automation Systems Simatic S7-200 family of programmable logic controllers. Siemens AG Automation and Drives Industrial Automation Systems Simatic S7-200 will be written as S7-200 for the remainder of the book. Information is provided to help the reader get and operate an inexpensive S7-200 programmable logic controller, the CPU 222, associated hardware, and software. Examples with ladder diagram programs and circuit diagrams are provided to demonstrate different S7-200 capabilities. Information is also provided to relate the CPU 222 to other programmable logic controllers. The person completing the examples will be able to write useful programs for the S7-200 controllers.

This book is written in the same format as my earlier books, *Programmable Logic Controller (PLC) Tutorial- Circuits and programs for Allen-Bradley MicroLogix and SLC 500 programmable controllers*, ISBN 0-9659446-6-2 and *Programmable Logic Controller (PLC) Tutorial, GE Fanuc- Circuits and programs for GE Fanuc VersaMax Nano and Micro programmable controllers*, ISBN 978-0-9659446-7-0.

To most people, a programmable controller, programmable logic controller, and PLC are the same thing. It should be noted, however, that PLC is a registered trademark that Allen-Bradley, a division of Rockwell Automation, uses to describe one of its lines of programmable logic controllers.

PLCs are simply special purpose computers that control electrically operated processes. The processes might be in chemical plants, steel mills, or other types of industrial plants that need precise control. Often PLCs do jobs that were formerly done by networks of relays and/or by teams of human operators.

Programmable controllers are more flexible, more rugged, easier to reprogram, and less expensive than all but the simplest relay logic systems. Also, PLCs can receive more types of inputs than relays. For example, some can receive input data from barcode scanners. Programmable controller knowledge is useful for maintenance technicians, plant engineering personnel, technologists, and engineers.

Any process controlled by a PLC has the following:
1) The process being controlled.
2) Input devices such as switches, sensors, or push buttons.
3) Input modules that act as a protective boundary and convert signals from the input devices to a form that can be used by the PLC's central processing unit (CPU), communication, and memory.

4) The PLC's CPU, communication, memory, and power supply.
5) A software program.
6) Output modules that act as a protective boundary and convert the CPU's output to a form that can operate external devices.
7) External devices such as lights, solenoids, and motor starters.
8) An operator terminal for programming and monitoring the control system and process.

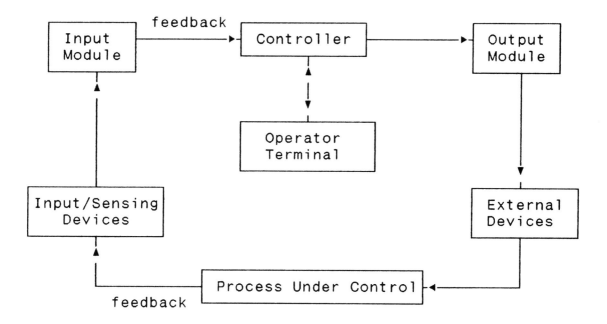

Figure I-1 Typical PLC configuration.

This book focuses on a 'starter' version of the S7-200 PLCs, the CPU 222. The CPU 222 is one of the smaller and less expensive S7-200 PLCs. The S7-200 family is programmed with its own software, "Step 7-Micro/WIN". The larger and more expensive Siemens S7-300 and S7-400 PLCs use a different Siemens software called "Step 7". "Step 7-Micro/WIN" and "Step 7" are not closely related, even though they both have "Step 7" in their names.

The person using this book should be able to read electrical circuit diagrams. He or she should be educated to or beyond the level of a two-year electrical technician's program.

The reader may need to contact Siemens personnel for help. Although efforts have been made to make this book up-to-date and accurate, the ever-increasing number of options and modifications available with the S7-200 PLCs and "Step 7-Micro/WIN" software may make some of this book's information dated.

Stephen P. Tubbs

1.0 SIEMENS FAMILIES OF PLCS

The main Siemens programmable logic controllers in order of capability and cost are:

1) S7-400. These are the most powerful of the SIMATIC PLCs. They are designed to handle large systems in manufacturing and process automation.
- The base unit consists of a backplane, a power supply, and a CPU.
- A diverse range of modules can be added at the backplane and at distributed structures.
- Available with standard, fail-safe, and fault-tolerant CPUs.
- Minimum bit execution time of .03 microseconds.
- Maximum main memory 20 MB.
- Maximum I/O address area 16384/16384 byte
- Programmable from a personal computer with "Step 7" software
- Programming languages LAD, FBD, IL, S7-Graph (SFC), S7-HiGraph, CFC
- Communication
    MPI
    PPI via CP
    PROFIBUS via CP
    PROFINET via CP

2) S7-300. These are less powerful and expensive than the S7-400s. They are the most popular of the S7 series and are designed to handle medium systems in manufacturing and process automation.
- The base unit consists of a backplane bus connector, a power supply, and a CPU.
- A diverse range of modules can be added to the backplane bus (the backplane bus is composed of cables and connectors connecting the modules, rather than card slots in a motherboard) and at distributed structures.
- Available with standard, compact, fail-safe, and technology CPUs.
- Minimum bit execution time of .01 microseconds.
- Maximum main memory 1.4 MB.
- Maximum I/O address area 8192/8192 byte
- Programmable from a personal computer with "Step 7" software
- Programming languages LAD, FBD, IL, S7-Graph (SFC), S7-HiGraph, CFC
- Communication
    MPI
    PPI via CP
    AS-Interface via CP
    PROFIBUS via CP
    PROFINET via CP
    PROFIDRIVE via CP

- PPI master/slave, MPI slave via an integrated port when using the Freeport mode
- Optional communications with expansion modules: Profibus DP Slave, AS-Interface master, Ethernet, and Internet.

4) CPU 222 base unit
- Dimensions 90 x 80 x 62 mm
- 8 digital inputs on base unit
- 6 digital outputs on base unit
- Accepts two expansion modules
- 78 maximum possible digital I/O with two expansion modules
- Input power 24 VDC or 120/230 VAC
- Digital outputs can be solid state (DC) or relay
- PPI master/slave, MPI slave via an integrated port when using the Freeport mode
- Optional communications with expansion modules: Profibus DP Slave, AS-Interface master, Ethernet, and Internet.

5) CPU 221 base unit
- Dimensions 90 x 80 x 62 mm
- 6 digital inputs on base unit
- 4 digital outputs on base unit
- Accepts no expansion modules
- Input power 24 VDC or 120/230 VAC
- Digital outputs can be solid state (DC) or relay
- PPI master/slave, MPI slave via an integrated port when using the Freeport mode

2.2 S7-200 OPTIONAL EXPANSION MODULES

CPUs 222 and larger accept optional expansion modules. There are 30 optional expansion modules that can be connected to the S7-200. These include 19 general purpose I/O digital & analog expansion modules, one RTD (Resistance Temperature Detector) input module, one thermocouple input module, one stepper or servo motor drive module, one weighing module, and seven communications modules. There are also five optional HMI (Human-Machine Interface) display and/or control panels. Details on these can be seen in the Siemens brochure, E20001-A1020-P272-X-7600 and the Siemens catalog, E86060-K4670-A111-A9-7600 (These can be found on the web addresses given in Section 2.3.2).

2.3 S7-200 PROGRAMMING SOFTWARE, "STEP 7-MICRO/WIN"

"Step 7-Micro/WIN" can be used to program all S7-200 PLCs.

IEC 1131 is the international standard for PLC programming languages. In accordance with IEC 1131, "Step 7-Micro/WIN" programs can be written as Ladder Logic Diagrams (LAD, similar to relay ladder connection diagrams), Function Block Diagrams (FBD, graphical dataflow programming), and Statement Lists (STL <also called IL, Instruction Lists>, text based programming).

Ladder Logic Diagram programming is the most popular for PLCs. It will be used in this book.

"Step 7-Micro/WIN" version 4.0 requires the following personal computer hardware and software:
- A personal computer (PC) running one of the operating systems:
    Microsoft Windows 2000 Service Pack 3 or later
    Microsoft Windows XP Home
    Microsoft Windows XP Professional
- At least 350 Mbytes of free hard disk space
- Small font setting and a minimum screen resolution of 1024x768 pixels
- Any mouse supported by Microsoft Windows
- For communicating with the S7-200 you will need one of the following:
    PC/PPI Cable to connect to a PC's USB port
    PC/PPI Cable to connect to a PC's serial communications port (PC COM1 or COM2)
    Communications processor (CP) card and multipoint interface (MPI) cable
    EM241 modem expansion module
    CP243-1 or CP243-1 IT Ethernet expansion module

At the time of writing, Siemens says that Microsoft Windows Vista may be compatible, although testing has not been done. Siemens plans to make available a new version (version 6.0) of "Step 7-Micro/WIN" that will be compatible with Microsoft Vista, XP, and 2000 in the September of 2007.

2.3.1 FREE "STEP 7-MICRO/WIN" 60-DAY SOFTWARE

Siemens offers a free fully functional demo version of "Step 7-Micro/WIN" version 4.0. This can be downloaded from their website, http://www.automation.siemens.com/_en/s7-200/support/tools_downloads/microwin.html. The demo can be used for 60 days and started a 100 times before it becomes inactive.

The installation instructions appear on screen as needed.

2.3.2 PURCHASED "STEP 7-MICRO/WIN" SOFTWARE

"Step 7-Micro/Win" software can be purchased directly from its own part number, 6ES7 810-2CC03-0YX0, or indirectly as part of a "starter kit".

The purchased software includes two CDs. The first is the Step 7-Micro/WIN software. The second contains system manuals, CP user manuals, tutorials, tips and tricks, and mechanical drawings.

To install simply follow the instructions that come with the software.

At the time of this book's writing, the CPU 222, "Step 7-Micro/WIN", PC/PPI cable, and manual were available for $199.00 in a special training promotion. The cost of the kit was much less than the cost of the kit's individual parts. Purchased separately, outside of the training promotion, the items of the kit had a list price of $479.00. I have heard that the Siemens sales staff often has some sort of sales promotion in progress. The buyer should read a current edition of the Siemens brochure, E20001-A1020-P272-X-7600 (on the web at https://www.click4business-supplies.siemens.de/Images_Artikel/E20001-A1020-P272-X-7600.PDF), and Siemens catalog, E86060-K4670-A111-A9-7600 (on the web at http://www1.siemens.cz/ad/current/layers/data/downloads/c003as/as_katalogy/as_katalogy_ST70/ST70_en.pdf). Then the buyer should work with his or her Siemens representative to find the best and most economical purchase.

2.3.3 SIEMEN'S "BASICS OF PLCS" COURSE AND S7-200 PROGRAMMABLE CONTROLLER SYSTEM MANUAL

Siemens has produced an onscreen tutorial titled, "Basics of PLCs" it can be found on the web at http://www.sea.siemens.com/step/templates/lessson.mason?plcs:1:1:1. It is also available in a paper copy as SEA Order No. STTM-EP10F-0605. Contact your Siemens representative to get a copy.

A paper copy of the *Simatic S7-200 Programmable Controller System Manual*, 6ES7 298-8FA24-8BH0, is included with the purchase of the training promotion kit.

3.0 S7-200 COMMUNICATIONS

It is not necessary to know much about the communications capabilities of the S7-200 to use it as a stand alone PLC. It is only necessary to know how to connect it through its EIA-485 (formerly called RS-485, EIA stands for Electronic Industries Association) sub D nine socket port by a cable adapter to a personal computer's serial or USB (Universal Serial Bus) port to download and upload "Step 7-Micro/WIN" programs and data. The logic programs and circuits in Chapter 7.0 will only use the EIA-485 port to connect to a personal computer. Other communications capabilities of the S7-200 will be generally discussed in this chapter, but will not be used elsewhere in the book.

The CPU 221 has only one possible communications port, its EIA-485 port. Through the EIA-485 port it can be connected to a personal computer for programming, system bus, display, operator panel, printer, modem, barcode reader, inverter, or other device. The CPU 221 will not accept expansion modules.

The CPU 222 PLCs and higher have one or two EIA-485 ports. From those, they all have at least the EIA-485 communications capabilities of the CPU 221. These CPUs also may be connected to expansion communication modules that connect them to the Industrial Ethernet, Internet, Profibus, or AS-Interface.

3.1 EIA-485 PORT

EIA-485 is a standard method for the configuration of data transmission networks. It specifies that data signal voltages are sent differentially over twisted wire pairs. It also specifies the electrical characteristics of the driver and the receiver. It does not specify the data protocol. The S7-200 data transmission rate through its EIA-485 ports varies from 1.2 to 187.5 kbits/second.

3.1.1 SYSTEM BUS DATA MODE

The S7-200s in the Freeport mode may connect their EIA-485 ports to a system bus. With a purely S7-200 network PPI (Point to Point Interface) protocol can be used. PPI can handle up to 126 stations. With a network that contains an S7-300, S7-400, or Simatic HMI (Human Machine Interface), S7-200s are integrated as MPI (Multi-Point Interface) slaves. MPI is a Siemens Automation programming protocol.

3.1.2 PROGRAMMING MODE

When a S7-200 is in the programming mode its data protocol through its EIA-485 is ASCII (American Standard Code II) or USS (Universal Simple Serial Interface, a Siemens registered trademark). With ASCII, it is possible to connect to modems, barcode scanners, personal computers, non-Siemens PLCs, etc. With USS protocol Siemens Sinamics inverters can be controlled.

3.1.3 MODBUS DATA MODE

S7-200s in the Freeport mode may also be connected through their EIA-485 ports to a Modbus RTU (Remote Terminal Unit) network. Modbus RTU is a transmission mode developed by the Modicon Company, but now open for use by all.

3.2 INDUSTRIAL ETHERNET

The original Ethernet protocol was first used on a LAN (Local Area Network) invented by the Xerox Corporation in the early 1970's. The original Ethernet network was made to connect a number of Xerox's computers to a single laser printer. The current Ethernet physical configuration and lower software layers are specified in the standard IEEE 802.3. Ethernet protocol is the most popular protocol for LANs.

The Industrial Ethernet standards specify more rugged hardware and software than those specified for the Office Ethernet. The physical environment is often more extreme in industrial settings and it is usually more important that industrial equipment not shut down on network faults. For example, a network operating an assembly line may endure greater temperature, humidity, and vibrations than a network in an office. Also the network running an assembly line may suffer serious problems with just a short interruption of service, but the network in an office could usually go off line for a minute or so with no serious problems.

There are two S7-200 Industrial Ethernet communications modules the CP 243-1 and the CP 243-1- IT. Both of these make remote programming, remote servicing and CPU to CPU communication possible.

For Industrial Ethernet networks, configure "STEP 7-Micro/WIN" to use TCP/IP protocol.

Siemens has produced a document on the Industrial Ethernet, "White Paper V1.0/2005 Simatic Net White Paper Basics of Industrial Ethernet". It can be seen on the web page, http://www.automation.siemens.com/download/internet/cache/3/1412996/pub/en/IE_Basics_WhitePaper_V1_0e.pdf.

3.3 INTERNET

CPU 222 and higher can be connected to the Internet. This is done by connecting the S7-200 to the communication module CP 243-1-IT. Then from the CP 243-1-IT there could be a connection to a DSL modem to the Internet. Through this connection the S7-200 can handle E-mail and send and receive files with FTP (File Transfer Protocol).

3.4 PROFIBUS DP

Profibus DP (Process Field Bus, Distributed Peripherals) is a remote I/O communication protocol specified by the European Standard EN 50170. It was developed in the early 1990s in Europe. Devices adhering to the standard are compatible regardless of their manufacturer.

With the EM 277 module, the CPU 222 and higher can be connected as a slave to Profibus DP.

More can be read about it in the brochure, *Profibus Siemens*, Order No. E86060-A4678-A171-A3-7600. The brochure is on the web at:
http://www.automation.siemens.co.uk/main/extra/literature/files/Automation%20Brochures/SIMATIC%20NET/Profibus/2006/e86060-a4678-a171-a3-7600%20-%20Profibus%20The%20Perfect%20Fit%20For%20The%20Process%20Industries.pdf

3.5 AS-INTERFACE

AS-Interface (Actuator-Sensor Interface) is a field networking system. It transfers machine-related digital and analog signals. It also acts as a universal interface between digital actuators and sensors and higher level controls.

It was introduced in the early 1990's. Since 1999, AS-Interface has been standardized according to EN 50295 and IEC 62026-2.

CPU 222 and higher can connect to the AS-Interface with the CP 243-2 AS-Interface master module.

More can be read about it in the brochure, *AS-Interface Siemens*, Order No. E20001-A550-P305-V1-7600. The brochure is on the web at https://www.automation.siemens.com/cd-static/material/info/e20001-a550-p305-v1-7600.pdf

3.6 S7-200 PC ACCESS

S7-200 PC Access is an OPC (Object Linking and Embedding for Process Control) server that provides a connection between a personal computer and up to eight S7-200s over any of the S7-200 protocols, PPI, MPI, modem, or Ethernet/IT CP. It operates through software that is loaded on to a personal computer and used along with Step 7-Micro/WIN.

More information on S7-200 PC Access can be found on
http://www.automation.siemens.com/_en/s7-200/support/tools_downloads/pcaccess.html

4.0 S7-200 CPU 222 "TEACHING SETUP"

4.1 TRAINING PROMOTION STARTER KIT

When this book was being written, Siemens was doing a training promotion. With this promotion, some Siemens PLC representatives were offering short training sessions and a starter kit for the very reasonable price of $199.

The starter kit contained:
1) CPU 222 S7-200 PLC Part No. 6ES7 212-1BB23-0XB0
2) Personal computer USB to CPU 222 EIA-485 cable adapter. Part No. 6ES7 901-3DB30-0XA0
3) Simulator switch circuit
4) Step 7-Micro/WIN V. 4.0 software Part No. 6ES7 810-2CC03-0YX0
5) S7-200 System Manual Part No. 6ES7 298-8FA24-8BH0

It was easy for me to set up this starter kit. I mounted the CPU 222 on a pine board base. The simulator switch circuit was connected to the CPU 222 and its switches numbered appropriately. With power off to the computer and CPU 222, the personal computer to CPU 222 cable was connected. A 115 VAC power cord was connected at L1 & N and the other end of the cord plugged into an electrical outlet. The personal computer was turned on. Then the CPU 222 was ready to be programmed and run.

A photograph and circuit diagram of the "teaching setup" follow:

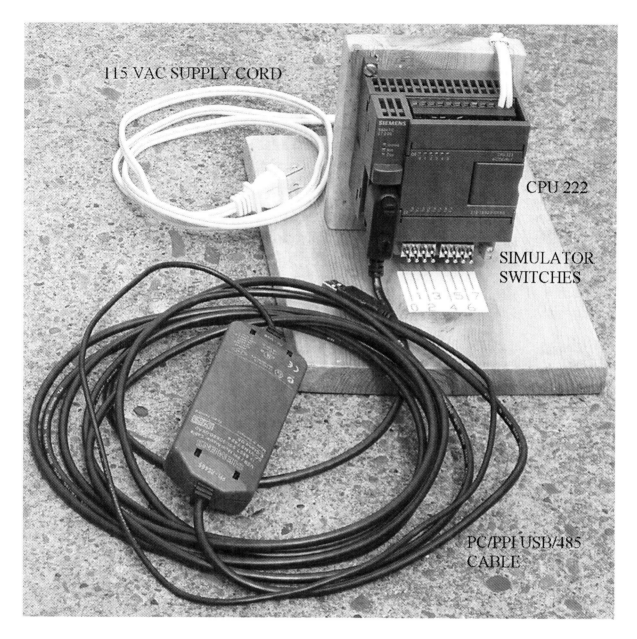

Figure 4-1 Photo of a complete "teaching setup" using the training promotion starter kit.

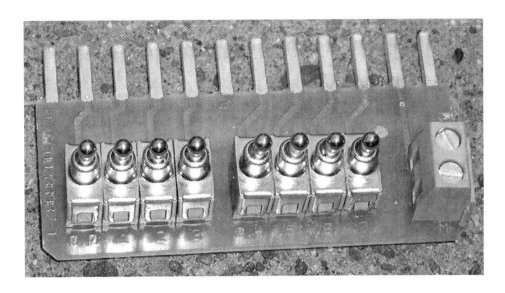

Figure 4-2 Photo of a the simulator switches supplied with the training promotion starter kit.

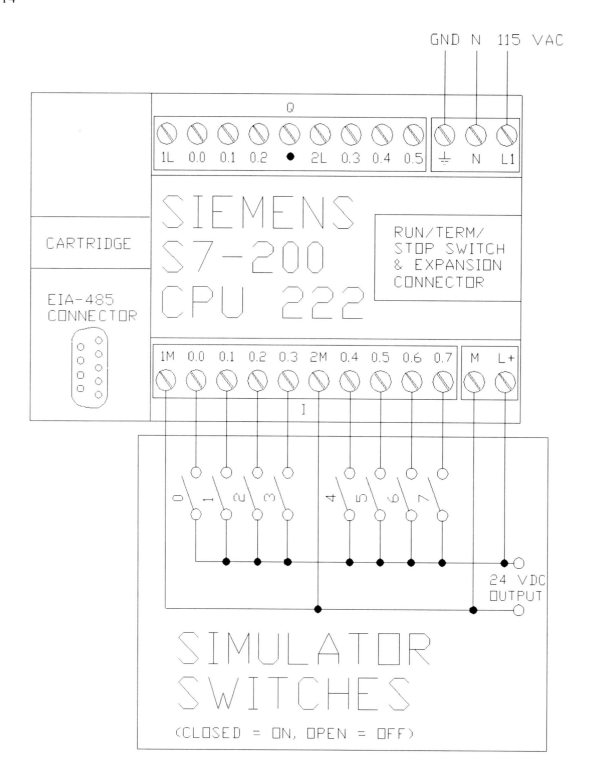

Figure 4-3 Schematic diagram of complete "teaching setup" using a "starter kit". The GND connection and the 24 VDC OUTPUT were not used.

4.2 ALTERNATIVE TO THE TRAINING PROMOTION STARTER KIT

If for some reason you can not get an economical starter kit you might want to build a "teaching setup" from parts. As mentioned above, the exercises of this book can be done with any of the S7-200 CPUs, including the least powerful and least expensive CPU 221.

If you are thinking of building from parts do not forget to add the expense of the Step 7-Micro/WIN programming software and the S7-200 to personal computer adapter cable.

In the books I wrote on the Allen-Bradley MicroLogix and GE Fanuc Nano I included purchase details and plans for a regulated 24 VDC supply and an input switch box. These are repeated here. Use them if you need them.

4.2.1 24 VDC SUPPLY

An already built 115 VAC/24 VDC power supply can be purchased to power a 24 VDC S7-200. A SOLA/HEVI-DUTY Power Supply capable of 1.3 A would work well. See www.grainger.com. Some of the SOLA power supplies are DIN rail mountable. They could be mounted on the same DIN rail as the S7-200.

It is also possible to build your own regulated DC supply. Building your own will give you a small savings and you may enjoy building it. Details on a home built power supply follow.

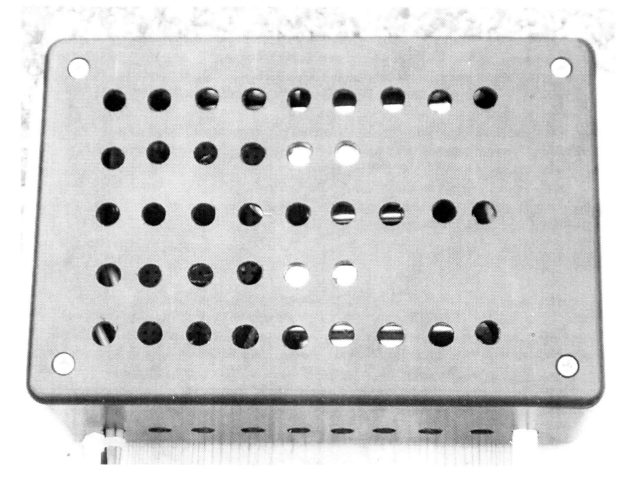

Figure 4-4 Photo of closed 24 VDC regulated power supply.

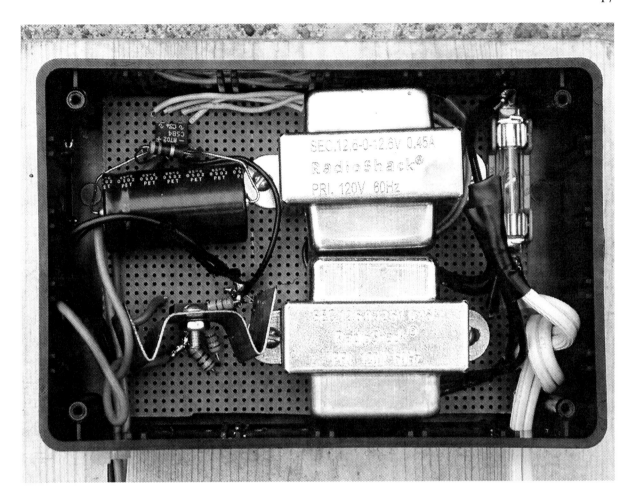

Figure 4-5 Photo of open 24 VDC regulated power supply.

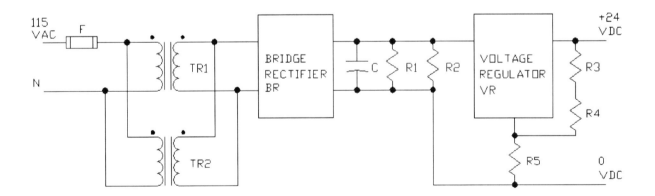

Figure 4-6 Schematic diagram of the 24 VDC regulated power supply.

Bill of materials:
    1) Plastic mounting box, 4" x 2.125" x 6".
    2) Perforation circuit board, 3.75" x 5".
    3) Fuse holder for 1" glass tube fuse.
    4) Fuse, 1" long glass tube, 250V, .5 A (F)
    5) Two 120/25.2V .45A output transformers. (TR1 & TR2)
    6) Electrolytic capacitor, 4700 microF, 35 VDC. (C)
    7) Bridge rectifier, 8702 CSB4, (Max. 400 PIV and max. 1 A average output on this device. A lower PIV higher current device could be used.) (BR)
    8) Two resistors, 4700 ohms +/- 10%. (R1 & R2)
    9) Resistor, 4700 ohms +/- 5%. (R3)
    10) Resistor, 150 ohms +/- 5%. (R4)
    11) Resistor, 100 ohms +/- 5%. (R5)
    12) Voltage regulator, LM317T PM19AN, (output voltage 1.3 to 37.5 VDC, maximum output current 1.5 ADC) (VR)
    13) Copper strip heat sink.

Construction notes:
    a) Solder all parts on to and through the perforated board.
    b) Holes were drilled in the plastic box to allow cooling air to go past the voltage regulator chip and transformer.
    c) All parts were purchased from Radio Shack.

4.2.2 SIMULATOR SWITCHES BOX

The simulator switches are mounted in a plastic box. The box shown has six digital inputs (8 are shown in the circuit of Figure 4-2, although six will do all exercises in this book). These would be connected the same way as shown in Figure 4-3.

Figure 4-7 Photo of closed input switch box.

Figure 4-8 Photo of open input switch box.

Bill of materials:
    1) Plastic box, 1.125" x 4" x 2".
    2) Switch numbers painted on with white correction fluid.
    3) Six 2-position switches, (SW0, SW1, SW2, SW3, SW4, SW5)

Construction note:
    All electrical parts were purchased from Radio Shack.

5.0 RELAY LADDER CONNECTION AND POWER DIAGRAMS

Relay ladder connection diagrams show electrical relays, contactor coils, and indicator lights in control circuits. Power diagrams show the electrical power wiring to motors and other power equipment.

Relay ladder connection diagrams look like ladders. They have conductors on the left and right that look like the rails of a ladder and they have circuits connecting those rails that look like rungs.

Figure 4-1 is a one rung ladder diagram. The push button and relay contacts "A" are the control elements. The relay coil "B" is controlled. The relay contacts "A" will close when relay "A" receives voltage. The coil for relay contacts "A" is not shown in Figure 4-1. When the push button is pressed and the relay contacts "A" are closed a conducting path is made to the relay coil "B". The connection of the relay coil "B" to voltage causes it to operate. Relay contacts for coil "B" are not shown in Figure 4-1.

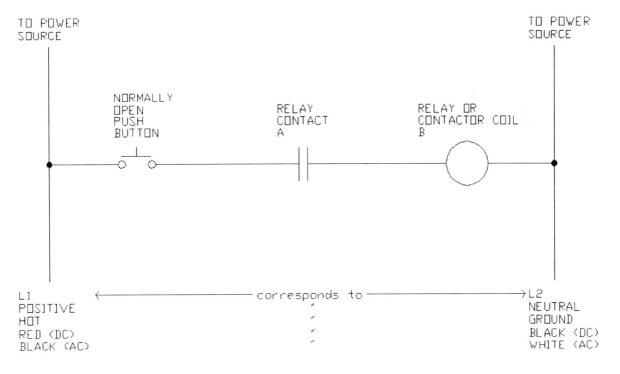

Figure 5-1 One rung ladder relay connection diagram.

More complicated relay ladder connection diagrams use more circuit symbols. Appendix 9.1, Relay Ladder and Power Diagram Symbols, shows commonly used symbols.

Following is an example of a control circuit that will run a three-phase induction motor in forward and reverse. The circuit is shown in a ladder relay connection diagram in Figure 5-2 and a power diagram in Figure 5-3. Notice that if the circuit breaker trips open or the power disconnect is opened the ladder relay connection diagram circuit will be de-energized and thereby reset to the "stop" condition. On many control circuits the FORWARD contactor and REVERSE contactor are mechanically interlocked to make it impossible for a malfunction to close both motor contactors at once.

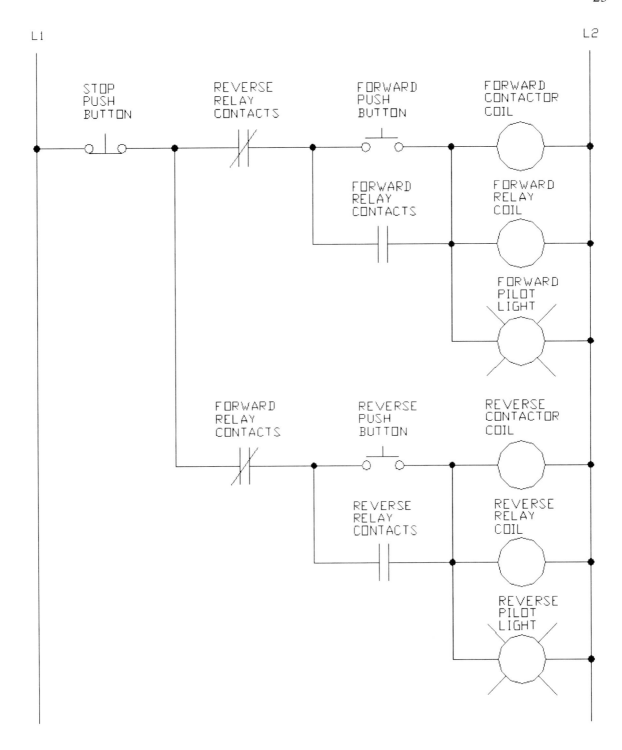

Figure 5-2 Forward and reverse motor control ladder relay connection diagram.

10) Left click OK to close the "Local Connection" and "Properties – PC/PPI cable(PPI)".

11) Double left click on the "Double-Click to Refresh" in the "Communications" window. When the connection is made an icon appears labeled "CPU 222 REL 02.01 Address:2".

12) The PC is now connected to the CPU 222.

6.4 PERSONAL COMPUTER INFORMATION THAT YOU MAY NEED

6.4.1 CHECKING HOW MANY COM AND USB PORTS ARE AVAILABLE ON YOUR MSWINDOWS XP PERSONAL COMPUTER

1) Left click mouse button on "START".
2) Right click on "MY COMPUTER".
3) Left click on "Properties".
4) Left click on "HARDWARE".
5) Left click on "Device Manager".
6) Double left click on "Ports (COM & LPT)".
7) Notice how many ports are available on your computer. (The only COM Port on my computer was COM 1.)
8) Double left click on "Universal Serial Bus Controllers".
9) Notice how many USB ports are available.

6.4.2 STOPPING ANOTHER PROGRAM'S USE OF A COM PORT

If you want "Step 7-Micro/WIN" to use COM 1 and another program has already grabbed away COM 1, "Step 7-Micro/WIN" will not connect.

On my computer it would have been necessary to stop RSLinx (an Allen-Bradley PLC to personal computer linking program) from using COM 1, if I wanted to use in for "Step 7-Micro/WIN". The RSLinx program that had been previously installed on my computer automatically grabs the COM 1 port when the computer boots up. You may have to stop another COM 1 grabbing program on your computer.

To stop RSLinx from using COM 1:
1) Left click on "START".
2) Left click on "CONTROL PANEL".
3) Double left click on "Administrative Tools".
4) Double left click on "Services".
5) Right click on "RSLinx".
6) Left click "Stop".
7) COM 1 is now available to be used by "Step 7-Micro/WIN".

6.5 USING THE S7-200 "RUN/TERM/STOP" SWITCH

For your computer to communicate with the CPU 222 the "Run/Term/Stop" switch must be in the Term (Terminal) or Stop position. The Run position allows the CPU 222 to run, but will not allow communication with your computer. The Term position is the best one to leave the CPU 222 on for this book, since it will allow the CPU 222 to both communicate and run.

7.0 EXAMPLE S7-200 CPU 222 LADDER AND POWER DIAGRAMS

The examples demonstrate the use of the S7-200 as a stand alone PLC and demonstrate important "Step 7-Micro/WIN" ladder diagram program instructions. For clarity, the example ladder diagram programs are short, sometimes using less than one page. Actual industrial ladder diagram programs would probably be longer.

Each example contains a ladder diagram program and CPU 222 circuit connection diagram. Most also contain a power circuit connection diagram. The student can write and operate the ladder diagram programs with only the "teaching setup". It is not necessary to build the CPU 222 circuits and power circuits. These circuits are only presented to clarify what the CPU 222 is designed to do in each example.

Ladder diagram programs differ from relay ladder connection diagrams in the way they are executed. Relay ladder connection circuits have voltage applied to all rungs at all times. (Note: "Step 7-Micro/WIN" uses the word "Network" instead of "Rung".) S7-200 and other PLC ladder diagram programs are executed one instruction at a time in scans. S7-200 and most other PLCs execute their instructions in a zigzag from top left, to top right, then down a Rung (Network) to the left side, then to the right, and so on. In some of the examples, the scan order is not important, in others the programs would not operate if the order of the ladder Rungs (Networks) were changed.

Notice the STOP push button, in the different examples, is normally closed. This way the CPU 222 will go into the stop mode when the push button opens. The advantage of doing a STOP this way is that if a wire in the STOP push button circuit were to break open the CPU 222 would stop. If the STOP circuit were controlled by a normally open push button a broken open wire would not stop the CPU 222 and if the STOP button were pressed the CPU 222 would not see a circuit change and would not stop the program.

The Section 7.1 example is most important because it shows the steps necessary to write, download and run ladder diagram programs. Most of these steps need to be repeated when doing the other examples.

7.1 CPU 222 LADDER AND POWER DIAGRAMS FOR THREE-PHASE INDUCTION MOTOR FORWARD AND REVERSE CONTROL

The relay logic scheme of Chapter 5.0 is redone using the simplest possible "Step 7-Micro/WIN" ladder diagram program and associated CPU 222 circuits and power circuits.

The new material demonstrated here is:
1) "Step 7-Micro/WIN".
2) "Normally Open" and "Normally Closed" ladder diagram program inputs.
3) Ladder diagram program "Outputs".
4) Putting addresses and descriptions in the ladder diagram program.
5) Compiling a ladder diagram program to check it.
6) Downloading a ladder diagram program.
7) Putting a S7-200 in the "RUN" mode.
8) Putting a running S7-200 in the "STOP" mode.
9) Observing a ladder diagram program's operation with "Step 7-Micro/WIN".
10) CPU 222 connection diagrams.

7.1.1 USING "STEP 7-MICRO/WIN" TO WRITE THE LADDER DIAGRAM PROGRAM ON A PERSONAL COMPUTER

7.1.1.1 Start "Step 7-Micro/WIN".

7.1.1.2 Left click on "File" in the upper left menu.

7.1.1.3 Left click on "New". Then left click on "Yes" or "No" as desired to save or not save the current project, if there was already a program open.

7.1.1.4 Under the "File" menu left click on "Save As". A screen appears that asks for the file name of your new project. Type in the file name, "Figure 7-4", select a directory to store it in and left click on "Save".

7.1.1.5 Check the PLC type by going to "PLC" on the upper menu, left clicking on it and then selecting "Type". The CPU 222 that was selected in Section 6.3, CPU 222, should already be there. If it is not, then select it now.

7.1.1.6 The program will show a "SIMATIC LAD" window that is ready to receive a ladder diagram program.

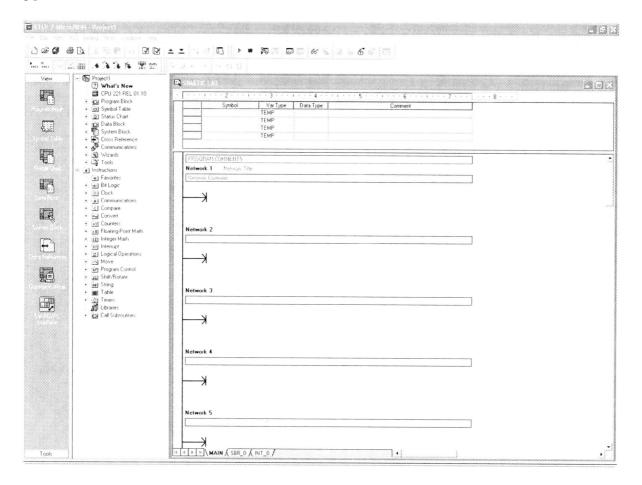

Figure 7-1 "Step 7-Micro/WIN" windows, before entering a ladder diagram program.

7.1.1.7 In the "Instruction Tree" window, the one headed "Figure 7-4 (C:\Program Files…" look under "Instructions". Left double click on "Bit Logic".

7.1.1.8 Drag the "Normally Open" contact symbol to the line below "Network 1".

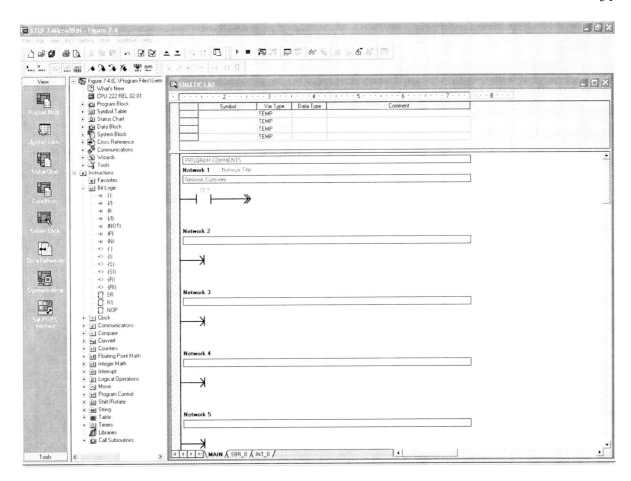

Figure 7-2 Ladder Diagram editor window with a "Normally-open" symbol added.

7.1.1.9 Left click, drag, and place the "Normally Open", "Normally Closed", "Output", "Line Up" and "Line Down" to make the diagram of Figure 7-3. It follows the design of the relay ladder connection diagram of Figure 5-2, although "Step 7-Micro/WIN" does not show the right ladder rail. Note the "Line Up" and "Line Down" symbols are in the toolbar above "SIMATIC LAD".

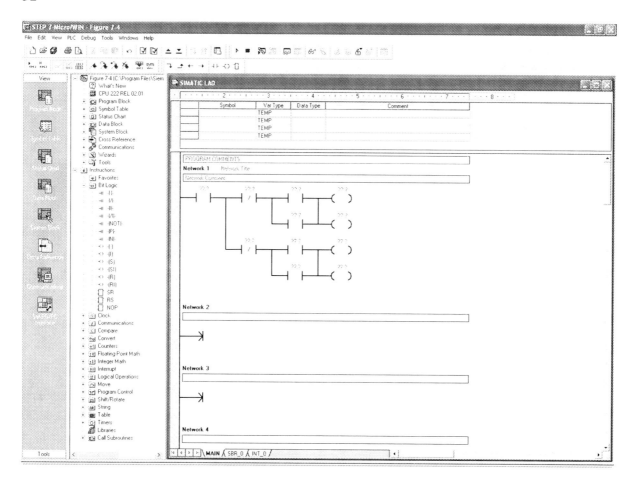

Figure 7-3 Ladder Diagram editor window with all needed instructions.

7.1.1.10 Left click on each "??.?" above the instruction symbols and type in the addresses shown in Figure 7-4. These are absolute addresses. Absolute addresses refer to real input points, real output points, or memory locations.

7.1.1.11 Type in a description for each address used in "Network 1" into the "Network Comment" block as shown in Figure 7-4. The descriptions will help the programmer understand and check the ladder program. The text in the "Network Comment" blocks is not used in the ladder program. You can type in whatever you want.

7.1.1.12 Rather than use the "Network Comment" block, it is possible to enter "symbolic" addresses of up to 13 capitalized characters. For example, variable I0.0 could be named STOPPB:I0.0. "Step 7-Micro/WIN" would put this in place of I0.0 above the "Normally Open" instruction in "Network 1" and would put it into a symbol table that could appear on the PC screen below the "Network 1" circuit. Using "symbolic" addresses may make programs easier to understand, but care is needed when using them. "Symbolic" addresses can be local to subroutines or global to a whole project. Program operation will be affected if the symbolic addresses are local when they should be global. To avoid global/local "symbolic" address problems and allow longer address descriptions, this book will only use absolute addressing and "Network Comment"s.

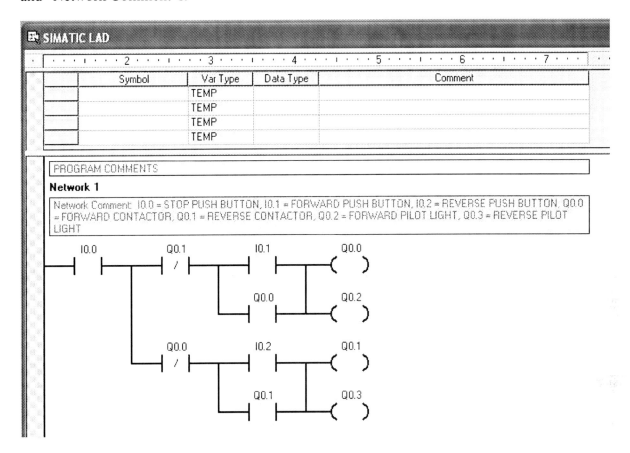

Figure 7-4 Ladder diagram program with addresses.

7.1.1.13 Save the program.

7.1.1.14 Check the program by left clicking "PLC" in the main toolbar and then left clicking "Compile All". Some other manufacturers call this "Validating". Any errors will appear at the bottom of the screen. When there are 0 errors, the program is ready to be downloaded and run.

7.1.1.15 The CPU 222 should be connected your PLC and "Step 7-Micro/WIN" configured, as was directed in Chapter 6.0. The "Run/Term/Stop" switch on the CPU 222 should be either on Term or Stop. Term would be the best choice. Once it is in that position, it can be left there for all the exercises in this book.

7.1.1.16 Download the program to the CPU 222 by going to "File" and left clicking "Download". If all is well a message will appear at the bottom of the screen saying "Download was Successful".

7.1.2 RUNNING AND STOPPING THE LADDER DIAGRAM PROGRAM

7.1.2.1 After the program has been downloaded it can be run two ways.

7.1.2.1.1 By switching the "Run/Term/Stop" switch to the "Run" position.

7.1.2.1.2 By switching the "Run/Term/Stop" switch to the "Term" position and controlling it from "Step 7-Micro/WIN". To control it from "Step 7-Micro/WIN" left click on "PLC" in the main toolbar. Then left click on "RUN" and left click on "Yes".

7.1.2.2 The program can be stopped three ways.

7.1.2.2.1 By removing power to the CPU 222. If this method is used the program will start automatically when power is reapplied if the "Run/Term/Stop" switch is in the "Run" position. If the "Run/Term/Stop" switch is in the "Term" position, the program will have to be restarted with "Step 7-Micro/WIN" via "PLC" on the main toolbar.

7.1.2.2.2 By switching the "Run/Term/Stop" switch to the "Stop" position.

7.1.2.2.3 By using "Step 7-Micro/WIN" to left click on "PLC" in the main toolbar and them left clicking on "STOP" and "Yes".

7.1.3 OBSERVING THE RUNNING LADDER DIAGRAM

7.1.3.1 The "Run/Term/Stop" switch should be in "Run" or "Term".

7.1.3.2 In "Step 7-Micro/WIN" left click on "Debug" in main toolbar.

7.1.3.3 Left click on "Start Program Status".

7.1.3.4 The ladder program diagram seen with "Step 7-Micro/WIN" should now show solid rectangles in the logically conducting contacts and energized outputs. This can be seen in Figure 7-5.

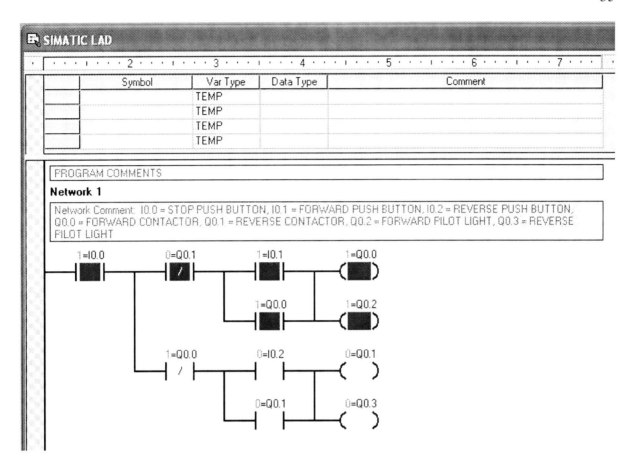

Figure 7-5 Ladder diagram program of operating CPU 222 with the forward contactor circuit active.

7.1.4 CPU 222 HARD-WIRED CIRCUIT

The CPU 222 hard-wired circuit can be made with or without compliance with National Electrical Manufacturing Association (NEMA) recommendations. NEMA recommends, *"When the operator is exposed to the machinery, such as loading or unloading a machine tool, or where the machine cycles automatically, consideration should be given to the use of an electromechanical override or other redundant means, independent of the controller, for starting or interrupting cycle"*. Following the NEMA recommendation, the STOP push button in our motor forward and reverse controller should be hardwired into the motor control circuit so that the motor will stop when the STOP push button is pushed regardless of whether the CPU 222 instructs it to or not. Two power diagrams will be presented, one with and one without compliance.

In both circuits CPU 222 power is received after the main circuit breaker and disconnect. This will cause the CPU 222 to reset to the "stop" condition soon after the main circuit breaker or disconnect opens.

7.1.4.1 CPU 222 circuit that does not comply with the NEMA safety recommendation.

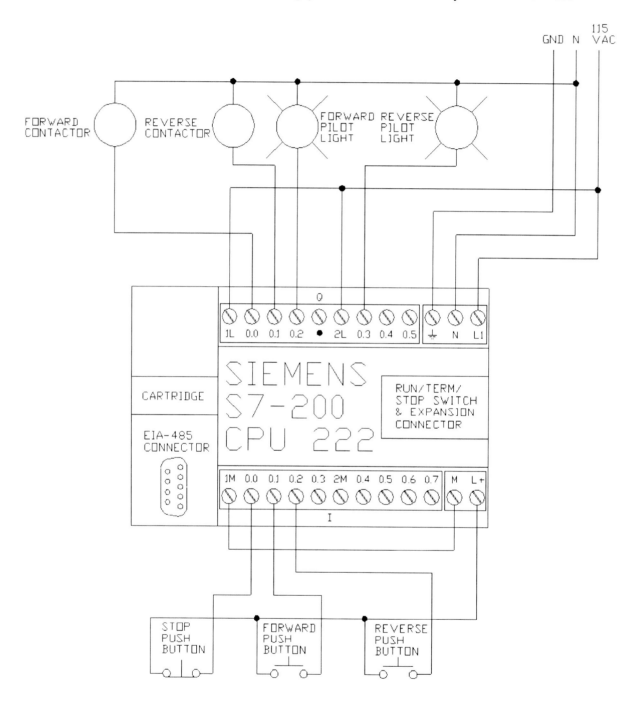

Figure 7-6 CPU 222 circuit without NEMA compliance.

7.1.4.2 CPU 222 circuit that complies with the NEMA safety recommendation.

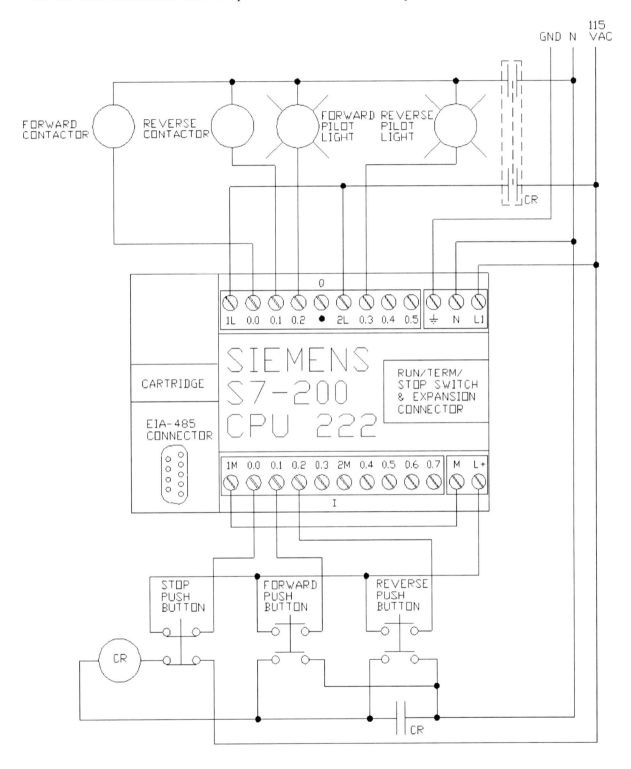

Figure 7-7 CPU 222 circuit with NEMA compliance.

7.1.5 POWER CIRCUIT

The same power circuit is used here as was used in Chapter 5.0 with the relay ladder connection diagram. See Figure 5-3, page 24.

7.1.6 WORTH OF USING A CPU 222 TO PERFORM FORWARD AND REVERSE SWITCHING

Is it sensible to use a CPU 222 to perform the forward and reverse switching of a motor rather than the simple relay ladder circuit of Chapter 5.0? No, it is not. Counting and comparing parts one sees that the CPU 222 only replaces two relays. Probably those two relays would be less expensive than the CPU 222 and the relays do not need programming software. The reason this CPU 222 circuit was put here was instructive, rather than practical. More complicated relay circuits involving more relays, counters, and timers would better justify the use of a CPU 222.

7.2 CPU 222 LADDER AND POWER DIAGRAMS FOR TIMED SEQUENTIAL STARTING OF TWO THREE-PHASE INDUCTION MOTORS

In this example, the cutting oil pump motor on a milling machine is controlled so that it starts and runs before and after the cutter motor operates. This assures that no cutting is done with an un-lubricated and un-cooled cutting tool. The NEMA recommended hardwired STOP push button is not used here or in following Sections, to allow us to focus our learning on the CPU 222 operation.

The new material demonstrated here is:
1) Internal "logical" contacts and coils.
2) F1 Help key.
3) "TON", On-Delay Timer, instruction.
4) Output "Set" and "Reset" instructions.

"Step 7-Micro/WIN" has internal outputs and contacts that can use M bit addresses. M addresses do not refer to physical inputs and outputs as I and Q addresses do. The M addresses are only part of the ladder diagram program.

Pressing the F1 key on your PC will provide help information about highlighted "Step 7-Micro/WIN" instructions and items.

"TON" is an instruction that logically goes on after it has received power for a time equal to the product of its PT (Preset Time) and the thousandths of a second multiplier next to ms. "TON" can be found in the "Instruction Tree" under "Timers". Drag it from there to the ladder diagram. When it is in the ladder diagram program, left click the "????" above the instruction. Enter a "T" and number. Different numbers after the T will put different millisecond (ms) time multipliers into the "TON" instruction. Use the F1 Help key on the highlighted "TON" instruction to see the selection of time multipliers.

An "Output Set" instruction looks like an ordinary "Output" instruction except for the "S" inside of it. However, an "Output Set" will not turn off by simply turning off its logical input. An "Output Set" stays on until both its logical input is off and its corresponding "Output Reset" has been turned on. The "Output Reset" instruction looks like an ordinary "Output" instruction with an "R" inside it. The "Output Set" and its corresponding "Output Reset" have numbers under them that tell how many bits should be set and reset.

Operation sequence with the ladder diagram program of Figure 7-8:
1) All systems off.
2) Circuit breaker disconnect switch closed. Power goes to the CPU 222.
3) The START push button is pressed. This resets the STOP, starts the oil pump motor, and starts the cutter motor "On Delay Timer", T37.

4) After a three second delay T37 starts the cutter motor.
5) Now the cutter motor and oil pump motor run simultaneously.
6) The STOP push button is pressed setting internal output M0.0 on.
7) The cutter motor has power removed immediately and the pump motor "On Delay Timer", T38, starts.
8) After a three second delay T38 removes power from the oil pump motor.

7.2.1 LADDER DIAGRAM PROGRAM

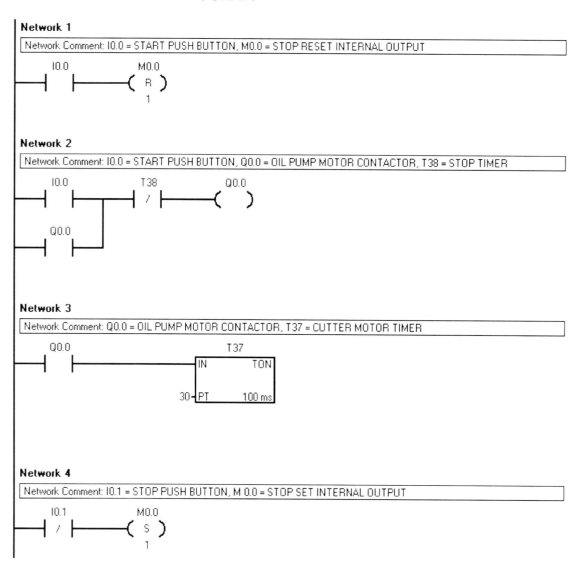

Continued

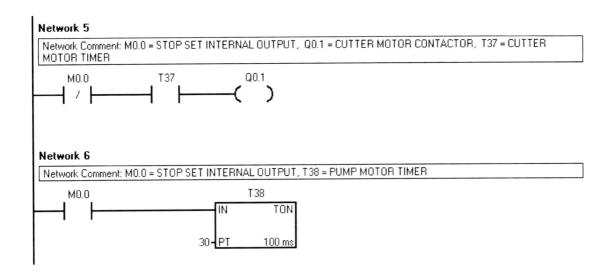

Figure 7-8 Ladder diagram program for the milling machine oil pump motor and drive motor controller.

7.2.2 CPU 222 CIRCUIT

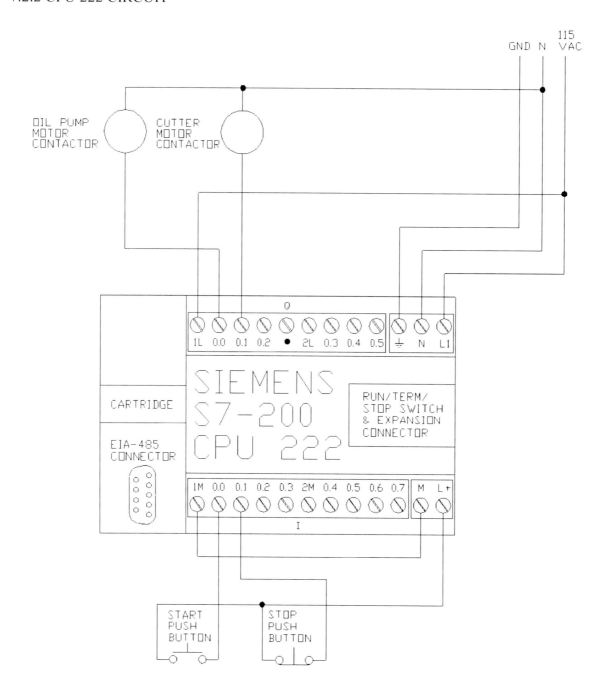

Figure 7-9 CPU 222 circuit for the milling machine oil pump motor and cutter motor controller.

7.2.3 POWER CIRCUIT

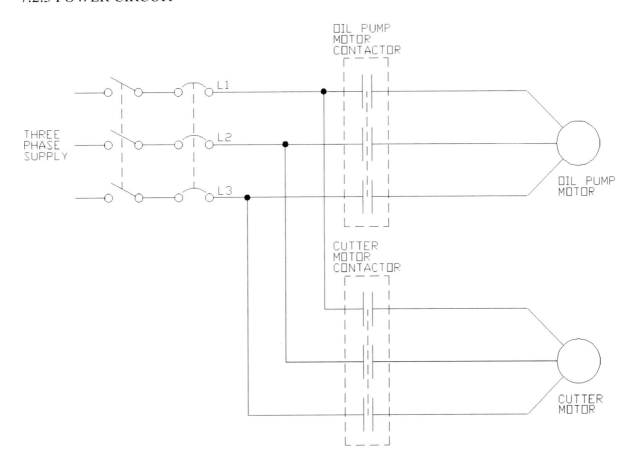

Figure 7-10 Oil pump motor and cutter motor power diagram.

7.3 CPU 222 LADDER AND POWER DIAGRAMS FOR A COUNTER-CONTROLLED THREE-PHASE INDUCTION MOTOR POWERED BOTTLE PUSHER

In this example, bottles on a packaging machine conveyor belt are pushed off to one side after six have passed. The pusher is made to move by a three-phase induction motor. After the pusher has gone to its push limit the motor reverses and the pusher returns to await six more bottles. This might be done in a bottling plant where the bottles are being dropped into boxes.

The new material demonstrated here is the "CTU", Count Up, instruction.

The "CTU" increments a number in a register for each "on" and "off" logical signal applied to its upper left input. When the R (Reset) is logically connected the "CTU" resets its count to zero and is ready to start counting again. The PV (Preset Value) is the number that the "CTU" is counting to. "CTU" is retentive; it will remember its count value even if the CPU 222 loses power.

The "CTU" can be found in the "Instruction Tree" under "Counters". Drag it from there to the ladder diagram. A number from 0 to 255 is placed after the C above the "CTU". This specifies the counter.

Operation sequence with the ladder diagram program of Figure 7-11:
1) All systems off.
2) Circuit breaker disconnect switch closed. Power goes to the CPU 222.
3) The pusher motor is at rest in the retracted position.
4) One bottle goes into the push area.
5) The BOTTLE COUNTER limit switch opens and closes once as the bottle passes.
6) The opening and closing of the BOTTLE COUNTER limit switch causes the "CTU" to increase its stored value by 1.
7) Five more bottles go into the push area. The BOTTLE COUNTER limit switch opens and closes five more times. This causes the stored number to increase by 5 to a total of 6.
8) Since C1 has reached its preset value of 6 the "CTU" is full and will now close the logical contact C1.
9) The closing of logical contact C1 causes the FORWARD TRAVEL contactor output to go on, pushing all six bottles off the conveyor belt.
10) After the motor has pushed to its limit, so as to open the FORWARD TRAVEL limit switch, the FORWARD TRAVEL contactor opens and the REAR TRAVEL contactor closes. This puts the motor in reverse and draws back the pusher.
11) When the pusher has returned to its retracted position the REAR TRAVEL limit switch opens, the motor shuts off, and the counter is reset.
12) The motor and counter are now ready to begin again.

7.3.1 LADDER DIAGRAM PROGRAM

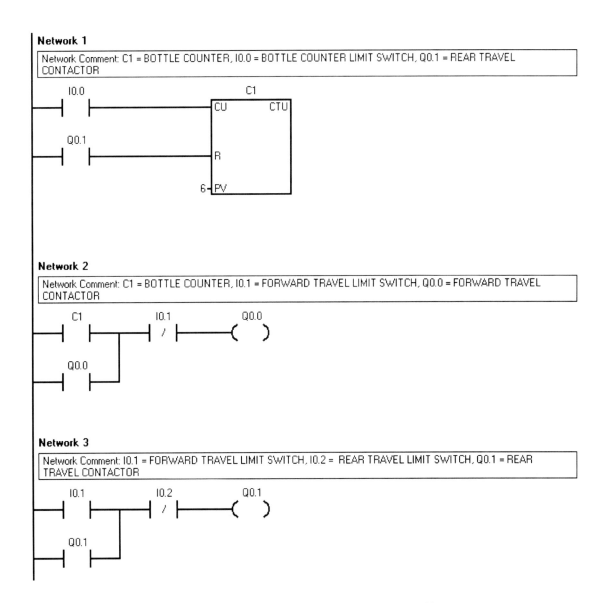

Figure 7-11 Ladder diagram program for the bottle pusher controller.

7.3.2 CPU 222 CIRCUIT

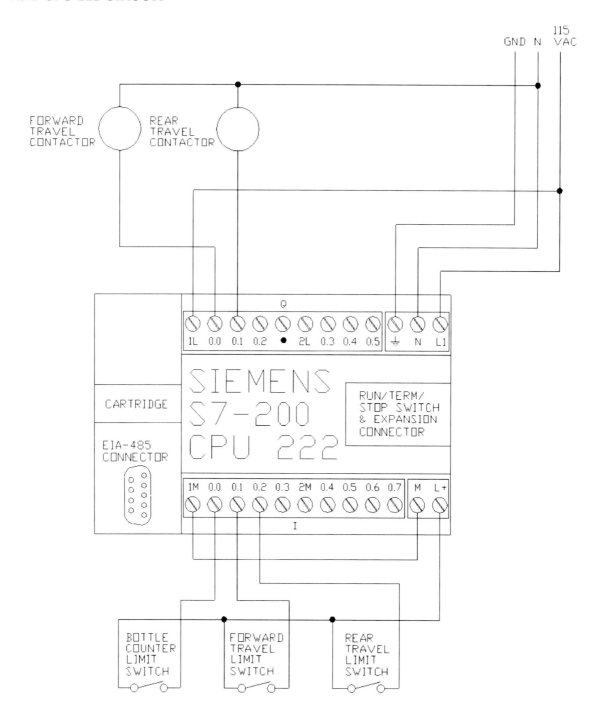

Figure 7-12 CPU 222 circuit for the bottle pusher controller.

7.3.3 POWER CIRCUIT

The same power circuit is used here as was used in Chapter 5.0 with the relay ladder connection diagram. See Figure 5-3, page 24.

7.4 CPU 222 LADDER AND POWER DIAGRAMS FOR A COUNTER-CONTROLLED THREE-PHASE INDUCTION MOTOR WITH A SETABLE COUNTER

This example is the same as 7.3 except that a switch has been added that allows an operator to change the counter from six to eight bottles.

The new material demonstrated here is:
1) The "MOV_B", Move Byte instruction. In this example, "MOV_B" moves either a 6 or 8 into the AC1 accumulator.
2) AC_, memory area. The accumulator memory areas AC0, AC1, AC2, and AC3 are 32-bit memory areas that data can be written to or read from. They can be used to pass program parameters or intermediate calculation values.

Operation sequence with the ladder diagram program of Figure 7-13:
1) All systems off.
2) Circuit breaker disconnect switch closed. Power goes to the CPU 222.
3) If the I0.3, 6 BOTTLES ON 8, BOTTLES OFF selector switch, is closed, "MOV_B" places 6 in AC1. Then the operation is identical to that in Section 7.3.
4) If the I0.3, 6 BOTTLES ON, 8 BOTTLES OFF selector switch, is open "MOV_B" places 8 in AC1.
5) The program will now count out 8 bottles rather than 6, but will otherwise be the same as that in Section 7.3.

7.4.1 LADDER DIAGRAM PROGRAM

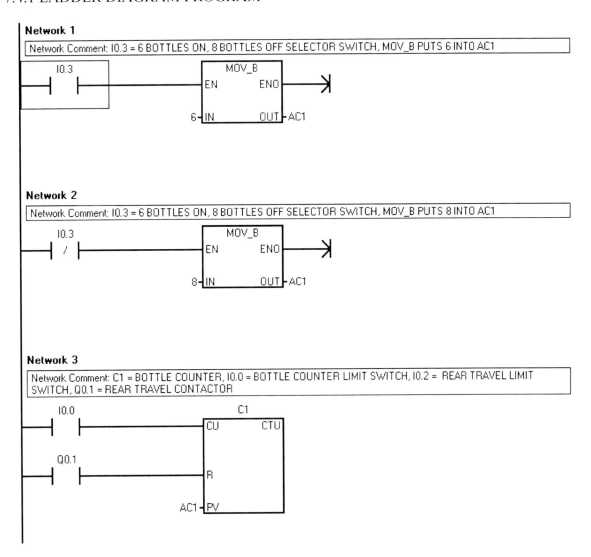

Continued

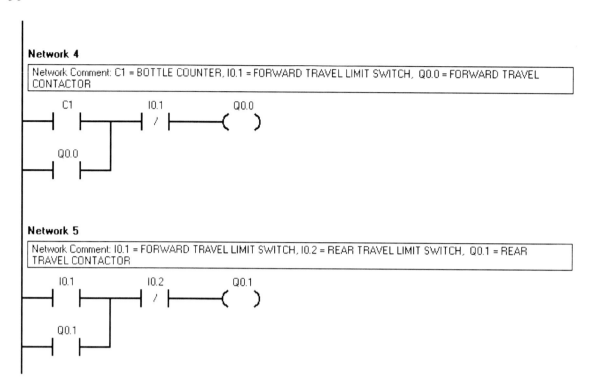

Figure 7-13 Ladder diagram program for the bottle pusher controller with a settable counter.

7.4.2 CPU 222 CIRCUIT

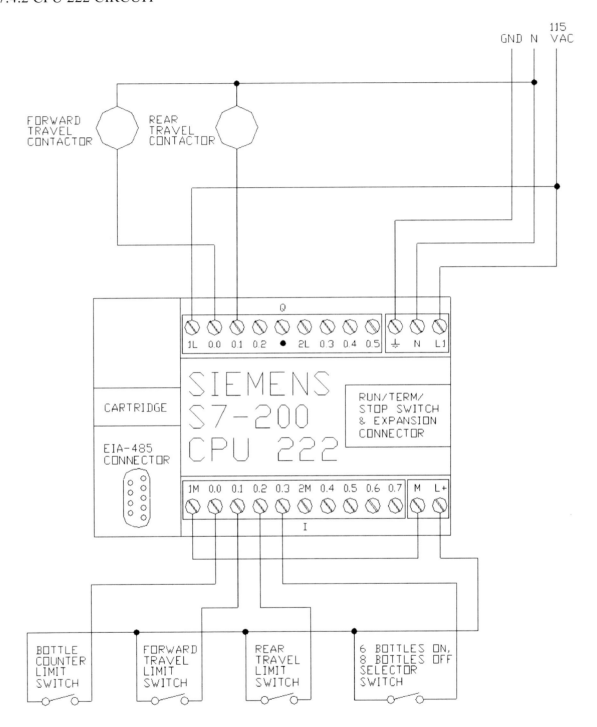

Figure 7-14 CPU 222 circuit for the bottle pusher controller with a settable counter.

7.4.3 POWER CIRCUIT

The same power circuit is used here as was used in Chapter 5.0 with the relay ladder connection diagram. See Figure 5-3, page 24.

7.5 CPU 222 LADDER AND POWER DIAGRAMS FOR ANOTHER COUNTER-CONTROLLED THREE-PHASE INDUCTION MOTOR

In this example, bottles coming from an input conveyor belt are accumulated in a holding area. If less than six bottles are in the area, a three-phase induction motor powered input conveyor belt is started. Once six bottles are accumulated, the power to the input conveyor belt's induction motor is turned off.

There is a limit switch that closes momentarily every time a bottle comes into the holding area and another that closes momentarily every time a bottle goes out of the holding area.

This program does not control the output conveyor belt.

The new material demonstrated here is:
1) The "SUB_I", Subtract Integer, instruction.
2) VW___ memory area. VW___ are 16-bit memory areas that can be used similarly to the AC_ memory areas that were used in Section 7.4.
3) The "⊣<I⊢", Less Than Integer, instruction.
4) The "SM0.0" always-on logical contacts.

Operation sequence with the ladder diagram program of Figure 7-15:
1) There are no bottles in the holding area.
2) The input conveyor belt brings in bottles to the holding area. The BOTTLE IN COUNTER limit switch, closes and opens as each bottle enters. C1 temporarily stores the number of openings.
3) The output conveyor belt takes bottles out of the holding area. The BOTTLE OUT COUNTER limit switch closes and opens as each bottle leaves. C2 temporarily stores the number of openings.
4) The difference between the numbers in C1 and C2 is calculated.
5) The difference between the C1 and C2 replaces the number in C1 and the number 0 is placed in C2.
6) If the difference is less than 6 the input conveyor belt motor continues to receive power via the CONVEYOR MOTOR contactor. If it is 6 or greater the CONVEYOR MOTOR contactor shuts off the power and the input conveyor stops.
7) When desired the counters can be reset to zero by closing the RESET COUNTER TO ZERO switch.

7.5.1 LADDER DIAGRAM PROGRAM

Network 1

Network Comment: I0.0 = BOTTLE IN COUNTER LIMIT SWITCH, I0.2 = RESET COUNTER TO ZERO SWITCH, C1 = BOTTLE IN COUNTER

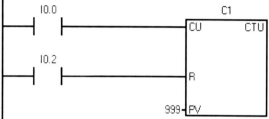

Network 2

Network Comment: I01 = BOTTLE OUT COUNTER LIMIT SWITCH, I0.2 = RESET COUNTER TO ZERO SWITCH, C2 = BOTTLE OUT COUNTER

Network 3

Network Comment: SM0.0 = ALWAYS ON DUMMY INSTRUCTION NEEDED TO MAKE SUB_I WORK, SUB_I = SUBTRACTS C1 COUNT BY C2 COUNT DIFFERENCE PLACED IN VW100

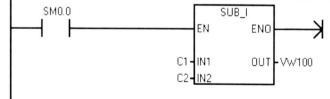

Continued

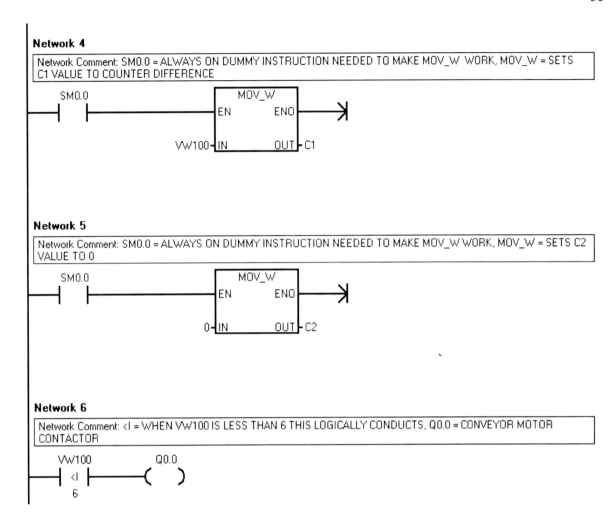

Figure 7-15 Ladder diagram program for the bottle accumulator.

7.5.2 CPU 222 CIRCUIT

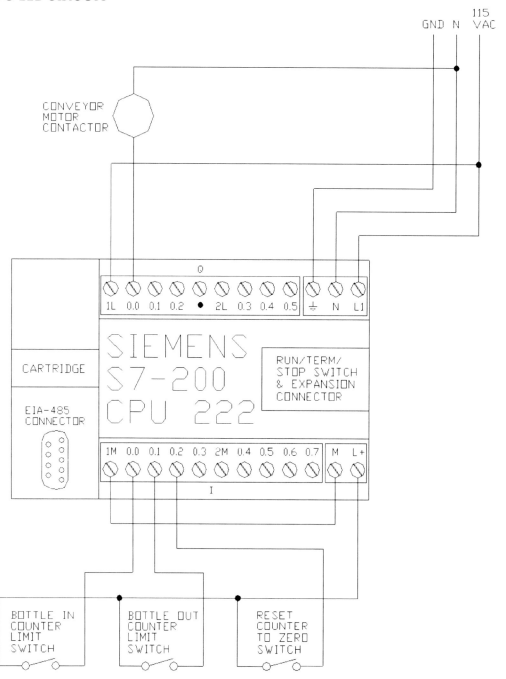

Figure 7-16 CPU 222 circuit for the bottle accumulator.

7.5.3 POWER CIRCUIT

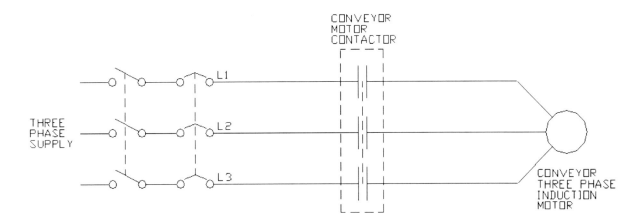

Figure 7-17 Bottle accumulator conveyor motor power diagram.

7.6 CPU 222 LADDER AND POWER DIAGRAMS FOR TIMED SEQUENTIAL STARTING OF TWO THREE-PHASE INDUCTION MOTORS WITH A PROGRAM CONTROL STOP

This is a modification of the example of Section 7.2. In Section 7.2 the oil pump motor cannot be immediately shutdown since there is a three second time delay after the STOP push button is pressed. This section's circuit includes an INSTANT STOP push button switch that instantly removes power to the milling machine's cutting motor and cutting oil pump motor by stopping the ladder program. As in Section 7.2 this will be done without the NEMA recommended hardwired STOP push button.

The new material demonstrated here is the program control "STOP" instruction. The program control "STOP" instruction turns off all outputs in a ladder diagram by putting a CPU 222 into the STOP mode.

The operation sequence of the ladder diagram program of Figure 7-18 without the INSTANT STOP push button being pressed, and with the push button contacts open, is the same as that of Figure 7-9 of Section 7.2.

Operation sequence of the ladder diagram program of Figure 7-18:
1) All systems off.
2) Circuit breaker disconnect switch closed. Power goes to the CPU 222.
3) The START push button is pressed.
4) The oil pump motor receives power and starts.
5) After a three second delay the main drive motor receives power and starts.
6) The main drive motor and oil pump motor run simultaneously.
7) The INSTANT STOP push button, is pressed.
8) The outputs controlling the main drive motor and oil pump motor have power removed immediately.
9) Letting the INSTANT STOP push button go back to its normally closed position makes the ladder diagram program available for normal operation again only when the CPU 222 goes back into the RUN mode. The CPU 222 can be put into the RUN mode again by switching its toggle switch to STOP and then back to RUN or leaving the toggle switch on TERM and putting the CPU 222 in the RUN mode with Step 7-Micro/WIN.

Industrial circuits that use the "STOP" instruction should have a programmer or technician reset the program after it has tripped.

7.6.1 LADDER DIAGRAM PROGRAM

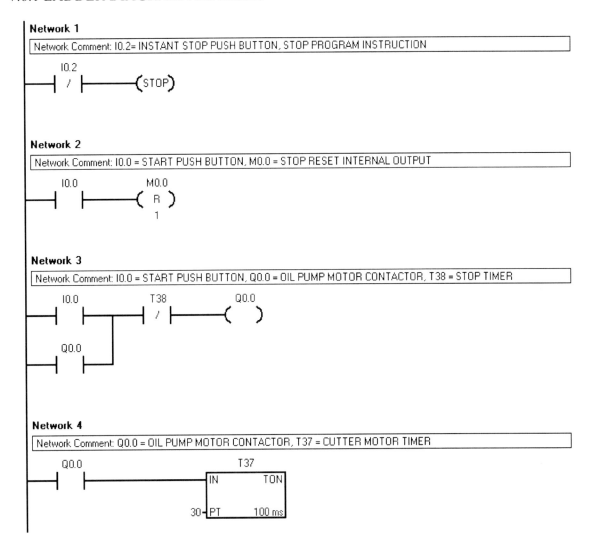

Continued

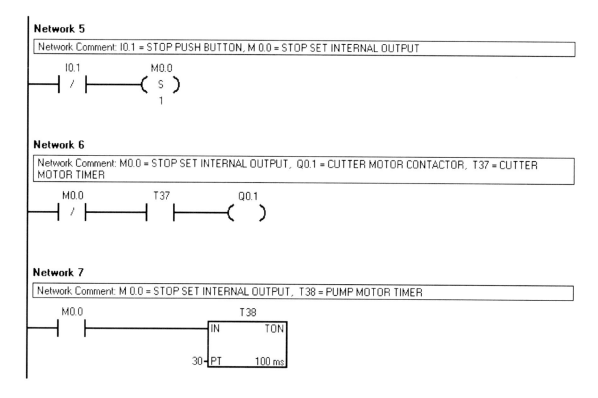

Figure 7-18 Ladder diagram program for the milling machine cutting oil motor/drive motor controller with a program control "STOP" instruction.

7.6.2 CPU 222 CIRCUIT

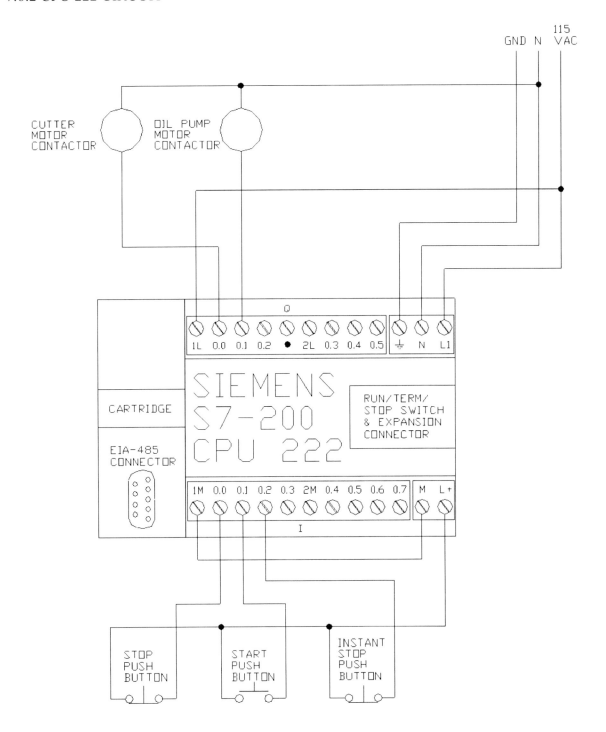

Figure 7-19 CPU 222 circuit for the milling machine cutting oil motor/drive motor controller with an instant stop

7.6.3 POWER CIRCUIT

The same power circuit is used here as was used in Section 7.2. See Figure 7-10, page 43.

7.7 CPU 222 LADDER AND POWER DIAGRAMS FOR A RETENTIVE TIMER CONTROLLED MACHINE

In this example, the CPU 222 stores a machine's total operating time. After the machine has operated for a set period of time, a warning pilot light goes on and the machine is automatically shutdown.

The new material demonstrated here is:
1) The "TONR", Retentive On-Delay Timer, instruction. It is retentive, remembering elapsed time, even when its activating input is off or the CPU 222 has its power removed.
2) The "cascading" of a timer with a counter.

Operation sequence with the ladder diagram program of Figure 7-20:
1) All systems off.
2) Power goes to the CPU 222.
3) The "CTU", Count Up Counter, and "TONR", Retentive On-Delay Timer, have the values from when the program was last run.
4) The MACHINE RUN push button is pressed, starting the machine through the MACHINE RUN contactor.
5) When the machine starts the "TONR" starts. It counts in 100 millisecond increments to a maximum of 5.0 seconds. The "TONR" time is "cascaded" through a "CTU" that effectively multiplies the time by 2. In the ladder diagram program this makes a resultant 5.0 x 2 = 10.0 second timer.
6) Cascading greatly extends the timing range of the CPU 222. The "TONR" is capable of counting to a maximum of 3,276.7 seconds, which equals only 54.6 minutes. With the "CTU" set to its maximum of 32,767 counts the maximum timing range is .1 x 32,767 x 32,767 seconds = 3.4 years. (If more time were needed more cascading could be done by adding another "CTU" to the ladder diagram program.)
7) When the "CTU" has reached its preset value, the machine is shut down and a MAINTENANCE NEEDED pilot light, goes on.
8) A RESET MACHINE RUN TIME push button, can be pressed. This turns off the MAINTENANCE NEEDED pilot light and puts the CPU 222 in a ready to start condition with zero time stored.

7.7.1 LADDER DIAGRAM PROGRAM

Network 1

Network Comment: I0.0 = MACHINE RUN PUSH BUTTON, I0.1 = STOP PUSH BUTTON, Q0.0 = MACHINE RUN CONTACTOR, Q0.1 = MAINTENANCE NEEDED PILOT LIGHT

```
   I0.0      I0.1      Q0.1      Q0.0
───┤├───┬───┤├───────┤├───────┤/├───────( )───
        │
   Q0.0 │
───┤├───┘
```

Network 2

Network Comment: Q0.0 = MACHINE RUN CONTACTOR, T5 TIMER = RETENTIVE TIMER, TURNS ON AT 5 SECONDS

```
   Q0.0              T5
───┤├────────────┤IN    TONR├
              50─┤PT   100 ms│
```

Network 3

Network Comment: C1 = COUNTS TO 2 AND TURNS ON, I0.2 = RESET MACHINE RUN TIME PUSH BUTTON, T5 CONTACT = CLOSES AFTER 5 SECONDS

```
   T5                C1
───┤├────────────┤CU     CTU├
                 │          │
   I0.2          │          │
───┤├────────────┤R         │
               2─┤PV        │
```

Continued

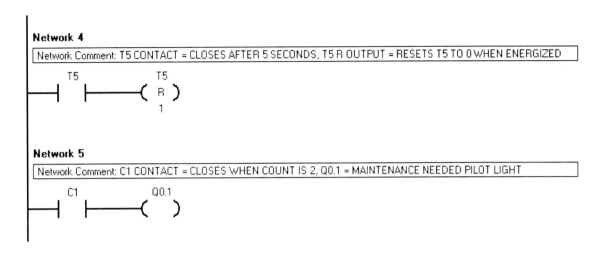

Figure 7-20 Ladder diagram program for the retentive timer controlled machine.

7.7.2 CPU 222 CIRCUIT

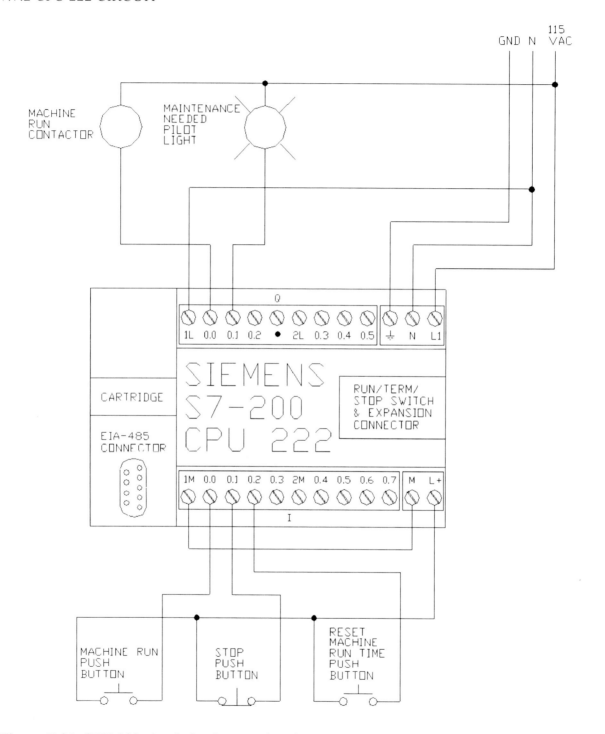

Figure 7-21 CPU 222 circuit for the retentive timer controlled machine.

7.8 CPU 222 LADDER AND POWER DIAGRAMS FOR A SYSTEM TO DETERMINE IF A BOTTLE COUNT RATE IS TOO LOW OR TOO HIGH

In this example, bottles riding on a conveyor belt are counted every 15 seconds. The number of bottles passing on the conveyor belt is compared to low and high limits. If the number is less than two bottles per 15 seconds a low rate output pilot light is energized. If the number is greater than or equal to four bottles per 15 seconds a high rate output pilot light is energized.

The new material demonstrated here is the "-| >=I |-", Greater Than or Equal Integer, instruction.

Operation sequence with the ladder diagram program of Figure 7-22:
1) All systems off.
2) Power goes to the CPU 222.
3) The BOTTLE COUNTER limit switch starts counting.
4) The BOTTLE COUNTING timer and DISPLAY RESET timer start timing in tenth of a second intervals.
5) After 14 seconds the DISPLAY RESET timer resets the LOW and HIGH pilot light outputs, turning both off.
6) After one more second the BOTTLE COUNTING timer causes the number of counts accumulated by the BOTTLE COUNTER limit switch, to be compared to the low limit of less than 2 and the high limit of 4 or more. The comparisons are done by the "Less Than Integer", and "Greater Than or Equal Integer" instructions. If the counts are less than 2 the LOW RATE pilot light, is turned on. If the counts are greater than or equal to four the HIGH RATE pilot light, is turned on.
7) The COUNTING TIMER limit switch causes the counter and timers to be reset to zero and the cycle, starting at 3), to begin again.

7.8.1 LADDER DIAGRAM PROGRAM

Network 1

Network Comment: C1 COUNTER = BOTTLE COUNTER, I0.0 = BOTTLE COUNTER LIMIT SWITCH, M0.2 = MASTER TIMER AND COUNTER RESET

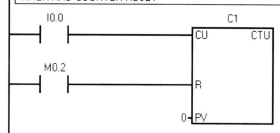

Network 2

Network Comment: SM0.0 = ALWAYS ON DUMMY CONTACTS NEEDED TO MAKE T37 WORK, T37 TIMER = BOTTLE COUNTING TIMER TURNS ON AT 15 SECONDS

Network 3

Network Comment: M0.2 = MASTER TIMER AND COUNTER RESET, T37 R OUTPUT = RESETS T37 TO 0 WHEN ENERGIZED

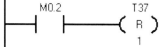

Network 4

Network Comment: SM0.0 = ALWAYS ON DUMMY CONTACTS NEEDED TO MAKE T38 WORK, T38 TIMER = DISPLAY RESET TIMER, TURNS ON AT 14 SECONDS

Continued

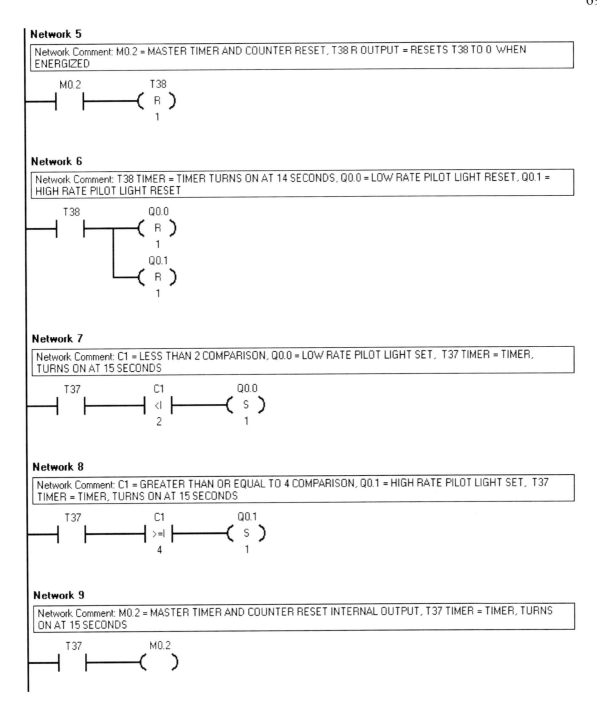

Figure 7-22 Ladder diagram program for the bottle rate checker.

7.8.2 CPU 222 CIRCUIT

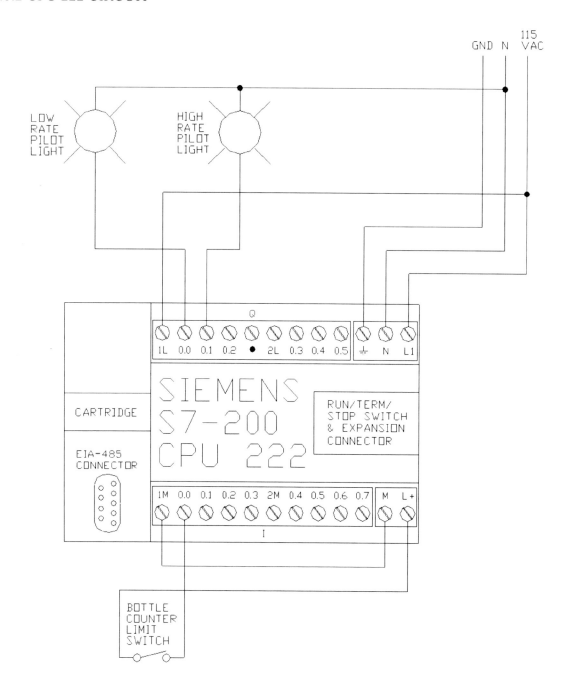

Figure 7-23 CPU 222 circuit for the bottle rate checker.

7.9 CPU 222 LADDER AND POWER DIAGRAMS FOR A SYSTEM USING BOTTLE COUNT RATE TO SELECT PRESET PUMP MOTOR INVERTER SPEEDS

In this example, two conveyor belts supply bottles. One supplies at a constant rate of three bottles/15 seconds. The other conveyor belt supplies bottles at a variable rate between zero and four bottles/15 seconds. The bottles of each conveyor belt are combined at the filling and capping machine. The combined bottle rate is used to select the pump motor speed for filling the bottles. Pump motor speed is set by the pump motor inverter output frequency. The bottle rate selects one of two preset speeds (output frequencies) on the pump motor inverter.

The new material demonstrated here is:
1) The "ADD_I", Add Integer, instruction.
2) The "-| >I |-", Greater Than Integer, instruction.
3) The "-| <=I |-", Less Than or Equal Integer, instruction.
4) The connection of CPU 222 outputs to inverter "preset speed" inputs.

Operation sequence with the ladder diagram program of Figure 7-24:
1) All systems off.
2) Power goes to the CPU 222.
3) The BOTTLE COUNTER limit switch starts counting.
4) The T38 counting timer, starts timing in tenth of a second intervals.
5) After 15 seconds the counts accumulated by the BOTTLE COUNTER limit switch, on the variable rate conveyor belt are added by the "Add Integer" instruction to those from the constant rate conveyor belt.
6) The total bottle count is compared to limits in the "Greater Than Integer" and "Less Than or Equal Integer" instructions. The appropriate inverter preset speed is turned on and the other off.
7) The C1 counter, and T38 counting timers, are both reset to zero.
8) The cycle, starting at 3), repeats.

7.9.2 CPU 222 CIRCUIT

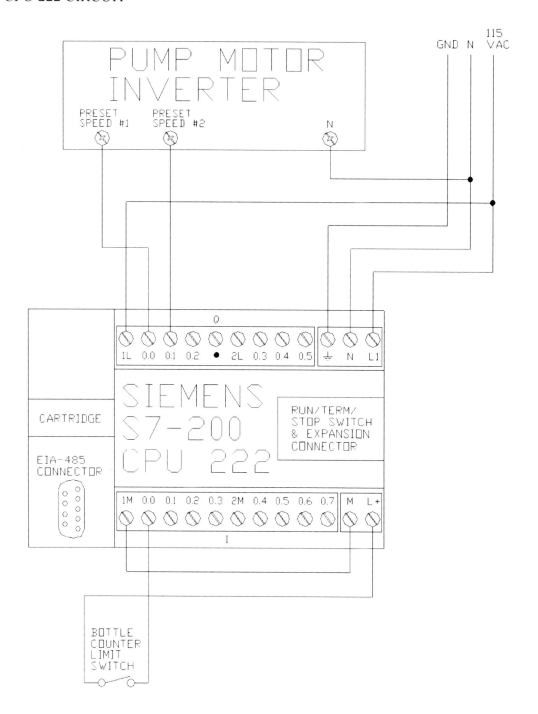

Figure 7-25 CPU 222 circuit for the bottle pump motor speed selector.

7.9.3 POWER CIRCUIT

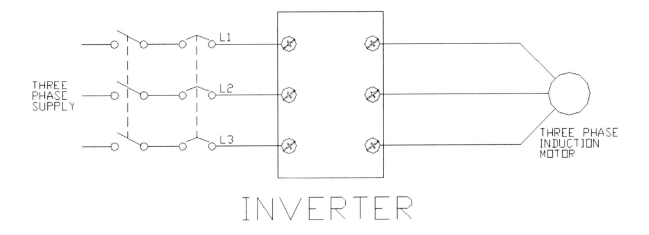

Figure 7-26 Bottle pump motor speed power diagram.

7.10 CPU 222 LADDER AND POWER DIAGRAMS FOR A CONVEYOR BELT PART PLACER

In this example, a conveyor belt is built with discrete trays. The conveyor belt trays may carry base parts or may be empty. A machine is to place mating parts on the base parts, but not on the empty trays. Five trays back from where the mating parts are added, there are two limit switches. One limit switch opens and closes each time a conveyor belt tray passes. The second limit switch opens and closes when a base part passes. The CPU 222 remembers the limit switch openings and closings. It uses the limit switch information it learned, from five tray locations back, to decide whether or not it should place a mating part.

The new material demonstrated here is:
1) The "⊣ P ⊢", Positive Transition instruction.
2) Making a variable binary rather than decimal.
3) The "SHL_B", Shift Left Byte instruction.
4) The "INC_B", Increment Byte instruction.

Numerical data is assumed by Step 7-Mico/WIN to be decimal. To have data stored otherwise, open the "Status Chart" in the "Navigation Bar". Type in the "Address" of the variable and select the "Format". In this case the variable should be binary, as shown in Figure 7-27.

	Address	Format	Current Value	New Value
1	VB100	Binary		
2		Signed		
3		Signed		
4		Signed		
5		Signed		

Figure 7-27 Making variable VB100 binary rather than decimal.

Operation sequence with the ladder diagram program of Figure 7-29:
1) The conveyor belt trays do not have base parts on them.
2) Power goes to the CPU 222.
3) The CPU 222 bit data for VB100 and V100.0, V100.1, V100.2, V100.3, V100.4, V100.5, V100.6, & V100.7 now looks like:

VB100	0000_0000							
V100.	7	6	5	4	3	2	1	0
STATE	0	0	0	0	0	0	0	0

VB100 is an 8-bit byte. Its individual bits are in the addresses V100.0, V100.1, V100.2, V100.3, V100.4, V100.5, V100.6, and V100.7. If VW100 had been used there would be 16 bits. If VD100 had been used there would be 32 bits.

4) The TRAY SENSING limit switch closes and opens as each tray goes past.

5) Each time the TRAY SENSING limit switch goes from open to closed it turns on "SHL_B" for one program scan.

6) The "SHL_B" instruction shifts the bits in VB100 one to the left. The "SHL_B" shifts one bit because 1 is written next to the N in "SHL_B". If 2 was next to the N instead of 1 then "SHL_B" would shift the bits two to the left. Bits on the far left are overwritten as bits from the right are shifted over them.

7) The BASE PART SENSING limit switch, connected to the input to "INC_B", determines the new bit value that goes into V100.0 (V100.0 is the rightmost bit in VB100). If the BASE PART SENSING limit switch is closed, as it would be when a base part is present, "INC_B" increments V100.0 by 1. If the BASE PART SENSING limit switch is open, a zero remains in V100.0.

8) Suppose there was a base part present, then the CPU 222 data would be:

VB100	0000_0001							
V100._	7	6	5	4	3	2	1	0
STATE	0	0	0	0	0	0	0	1

9) If on the next tray that passes there is also a base part, the CPU 222 data would be:

VB100	0000_0011							
V100._	7	6	5	4	3	2	1	0
STATE	0	0	0	0	0	0	1	1

10) If on the next tray that passes there is not a base part, the CPU 222 data would be:

VB100	0000_0110							
V100._	7	6	5	4	3	2	1	0
STATE	0	0	0	0	0	1	1	0

11) As each tray goes past, 0's and 1's will be shifted to the left until they disappear past V100.7.

12) The important bit for this program is at V100.5. When V100.5 is 1 then the MATING PART PLACING MACHINE places a mating part.

13) The values of the V100 and V100.5 can be observed during the ladder diagram program's operation. To see them, look at the "Status Chart" using the "Start Chart Status" feature under the main toolbar's "Debug". Figure 7-28 shows the "Status Chart" when the condition of Step 10) is occurring. The "2#" in the "Current Value" shows that the data is binary.

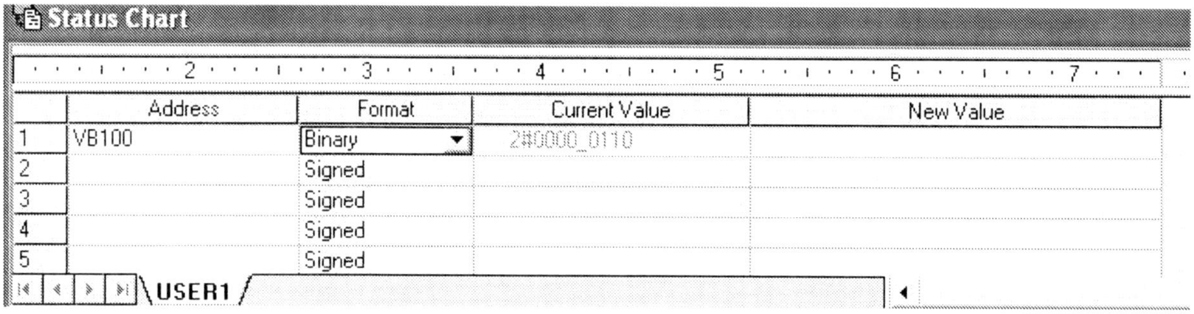

	Address	Format	Current Value	New Value
1	VB100	Binary	2#0000_0110	
2		Signed		
3		Signed		
4		Signed		
5		Signed		

Figure 7-28 Status Chart during program operation.

7.10.1 LADDER DIAGRAM PROGRAM

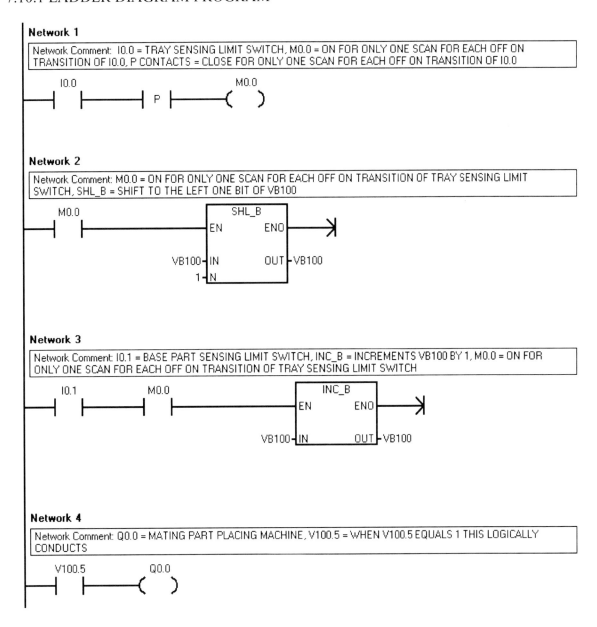

Figure 7-29 Ladder diagram program for the conveyor belt part placer.

7.10.2 CPU 222 CIRCUIT

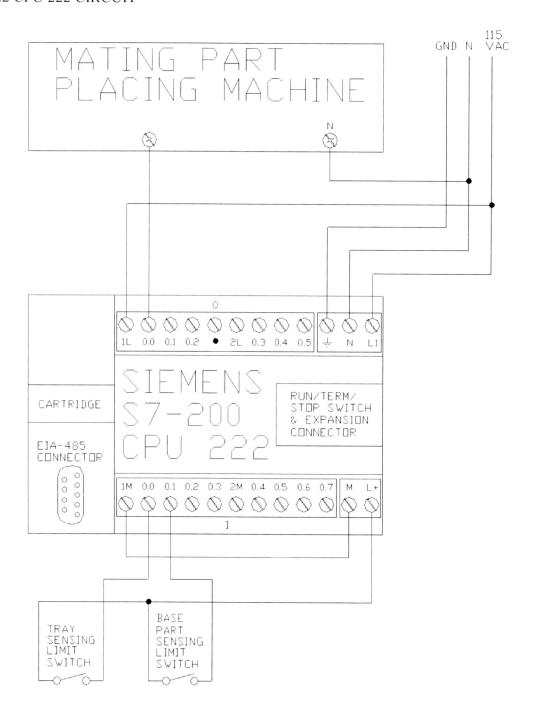

Figure 7-30 CPU 222 circuit for the conveyor belt part placer.

7.11 CPU 222 LADDER AND POWER DIAGRAMS FOR A CONVEYOR BELT SPEED CONTROLLER

In this example, three push buttons are used to start, set the speed, and stop a conveyor belt. A "ROL_B", Rotate Left Byte instruction, is the heart of the ladder diagram program. It is used much like a mechanical drum switch. In operation, first the START push button is pressed to put the conveyor belt in the run mode. Then the SPEED SET push button is pressed until the desired speed is set. At anytime, pressing the STOP push button puts the controller in the stop mode. Once in the stop mode, the START push button has to be pressed to re-start the conveyor.

The new material demonstrated here is:
1) "ROL_B", Rotate Left Byte, instruction.
2) The order of the ladder diagram program scan as a programming technique.
3) Manual entry of a V100.0 value in the "Status Chart" tables.

For the ladder diagram program to operate properly, the value of V100.0 is initially set to 1. This could have been written into the ladder diagram program. However, here it has been entered into the "Status Chart" table after the ladder diagram program was written and downloaded. The procedure for entering the value is:

1) Write and download the ladder diagram program of Figure 7-32 into the CPU 222.
2) Open the "Status chart" window.
3) Enter the value 2#0000_0001 into the "New Value" column as shown in Figure 7-31.
4) Left click on the padlock in the upper toolbar. This padlock is called "Force". The value 2#0000_0001 will be moved into the "Current Value" column.
5) There will be a closed padlock next to the 2#0000_0001. This means the value will be kept as 2#0000_0001 regardless of the program. To make it so that the program can change it, left click on the open padlock (called "Unforce") in the upper toolbar. In this example, V100 needs to be changeable for the program to operate properly.

	Address	Format	Current Value	New Value
1	VB100	Binary		2#0000_0001
2		Signed		
3		Signed		

Figure 7-31 Status Chart with 2#0000_0001 entered.

Operation sequence with the ladder diagram program of Figure 7-32:
1) The conveyor belt is stopped.
2) Power goes to the CPU 222.
3) VB100 now looks like:

VB100	0000_0001						
V100.	7	6	5	4	3	2	1	0
STATE	0	0	0	0	0	0	0	1

VB100 is one 8-bit byte containing V100.0 to V100.7.

4) At this time, the conveyor is stopped, V100.0 equals 1, and CONVEYOR BELT MOTOR INVERTER STOP is on.

5) When the SPEED SET push button is pressed, "ROL_B" moves the 1 in V100.0 two spaces to the left to V100.2. The 2 is indicated next to N in the "ROL_B" instruction. This causes the conveyor to go to its lowest speed where CONVEYOR BELT MOTOR INVERTER LOW SPEED is on and the other outputs are off.

VB100	0000_0100						
V100.	7	6	5	4	3	2	1	0
STATE	0	0	0	0	0	1	0	0

6) When the SPEED SET push button is pressed again "ROL_B" moves the 1 in V100.2 two spaces to the left to V100.4. This causes the conveyor to go to its middle speed where CONVEYOR BELT MOTOR INVERTER MIDDLE SPEED is on and the other outputs are off.

VB100	0001_0000						
V100.	7	6	5	4	3	2	1	0
STATE	0	0	0	1	0	0	0	0

6) When the SPEED SET push button is pressed again "ROL_B" moves the 1 in V100.4 two spaces to the left to V100.6. This causes the conveyor to go to its high-speed where CONVEYOR BELT MOTOR INVERTER HIGH SPEED is on and the other outputs are off.

VB100	0100_0000							
V100._	7	6	5	4	3	2	1	0
STATE	0	1	0	0	0	0	0	0

8) When the SPEED SET push button is pressed again "ROL_B" moves the 1 in V100.6 one space to the left and around one space so that it rotates to V100.0. This causes the conveyor to stop where CONVEYOR BELT MOTOR INVERTER STOP is on and the other outputs are off. Once stopped, the START push button has to be pressed again to restart the conveyor.

VB100	0000_0001							
V100._	7	6	5	4	3	2	1	0
STATE	0	0	0	0	0	0	0	1

9) At any speed when the STOP push button is pressed, the "ROL_B" instruction will rotate itself to where V100.0 equals 1, the position where CONVEYOR BELT MOTOR INVERTER STOP is on. Once stopped, the START push button has to be pressed again to allow speeds to be increased.

10) A significant difference between relay ladder connection diagrams and ladder diagram programs is that the rungs on a relay ladder are energized simultaneously, but the rungs on a PLC ladder diagram program are enabled one at a time. The rungs on a PLC are scanned through. With Siemens PLCs, the rungs (called Networks by Siemens) are read and executed left to right across the rungs and then down to the left side of the next rung in a zigzag pattern.

The vertical order of the rungs is critical to the proper operation of this ladder diagram program. When the STOP push button is pressed, the program itself moves the 1 in VB100 to the left and then rotates it to the stop position of V100.0. It makes one move per program scan. When the program reaches the stopped position, it stops operating "ROL_B".

11) This program is more difficult to describe in detail than the previous ones. The reader would better understand it by making a copy(s) of it and then following through the copy with a pencil.

7.11.1 LADDER DIAGRAM PROGRAM

Network 1

Network Comment: I0.0 = START PUSH BUTTON, I0.1 = STOP PUSH BUTTON, M0.0 = RUN MODE

```
    I0.0         I0.1         M0.0
 ───┤ ├───┬─────┤ ├──────────( )───
           │
    M0.0   │
 ───┤ ├───┘
```

Network 2

Network Comment: I0.2 = SPEED SET PUSH BUTTON, M0.0 = RUN MODE, P CONTACTS = TURN ON ONCE PER SPEED SET BUTTON PRESS WHEN IN RUN MODE, M0.1 = TURN ON ONCE PER SPEED SET BUTTON PRESS WHEN IN RUN MODE

```
    I0.2         M0.0                    M0.1
 ───┤ ├─────────┤ ├─────────┤ P ├───────( )───
```

Network 3

Network Comment: I0.0 = START PUSH BUTTON, I0.1 = STOP PUSH BUTTON, M0.1 = TURN ON ONCE PER SPEED SET BUTTON PRESS WHEN IN RUN MODE, M0.2 = STOP MODE, V100.6 = HIGH SPEED

```
    I0.1                     M0.2
 ───┤/├───┬─────────────────( )───
          │
    M0.2  │   I0.0
 ───┤ ├───┼───┤/├───
          │
    M0.1  │  V100.6
 ───┤ ├───┴──┤ ├───
```

Network 4

Network Comment: M0.1 = TURN ON ONCE PER SPEED SET BUTTON PRESS WHEN IN RUN MODE, M0.2 = STOP MODE, M0.3 = STOP MODE, M0.4 = SPEED CHANGE

```
    M0.1                     M0.4
 ───┤ ├──────────────────────( )───
          │
    M0.2      M0.3
 ───┤ ├─────┤/├───
```

Continued

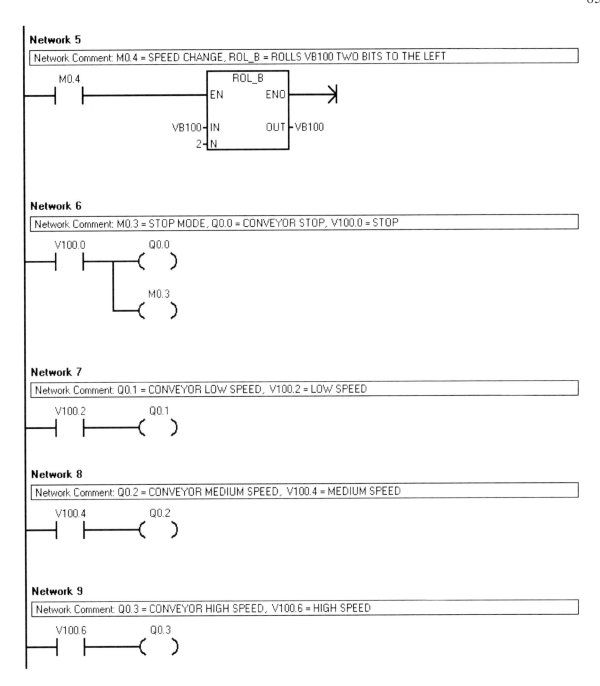

Figure 7-32 Ladder diagram program for the conveyor belt speed controller.

7.11.2 CPU 222 CIRCUIT

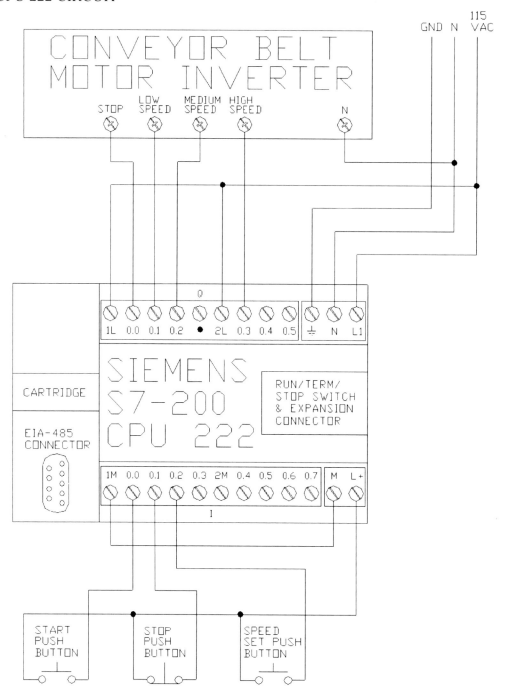

Figure 7-33 CPU 222 circuit for the conveyor belt speed controller.

7.11.3 POWER CIRCUIT

The same power circuit is used here as was used in Section 7.9. See Figure 7-26, page 75.

7.12 CPU 222 LADDER AND POWER DIAGRAMS FOR A SYSTEM THAT USES TABLE STORAGE AND FIFO TO SELECT SPRAY PAINT COLORS

In this example, the CPU 222 is used to control paint spray color in an assembly line painting booth. When assembly line items enter a holding area prior to the paint booth, a bar code reader reads the paint colors from bar codes attached to the items. The bar code reader closes the appropriate color switch for each item. The CPU 222 stores the color selection in a table. When an item leaves the holding area, the CPU 222 uses the table's first (oldest) stored color to select the appropriate color spray paint for the item. It then "forgets" that color.

The new material demonstrated here is:
1) The "MOV_W", Move Word, instruction.
2) The "AD_T_TBL", Add to Table, instruction.
3) The "FIFO", First In First Out, instruction.
4) The "─| ==I |─", Equal Integer, instruction.

A blank "Status Chart", showing only variable names, but not values, was created when the ladder program was written. After the program was downloaded into the CPU 222, the value of VW200 was Forced to 6 (The procedure for Forcing is in Section 7-11). The 6 in VW200 indicates that the table being used has six spaces for word storage. These spaces are in VW204, VW206, VW208, VW210, VW212, and V214. The six storage spaces have numerical values that were entered into the chart automatically by the operating ladder diagram program. Values can be observed during CPU operation by using the "Start Chart Status" under the main toolbar "Debug".

Figure 7-34 shows the "Status Chart" during CPU 222 operation with the code for red (1) stored in VW100 and in VW208. The VW400 has the code for red (1) in it. The VW204 has the code for green (100) that is ready to be loaded next into the VW400. VW206 has the code for yellow (10) stored in it. . The 3 in VW202 shows that there are now 3 values stored in the table, in VW204, VW206, and VW208. The values in VW210, VW212, and VW214 have no meaning to the ladder program, unless more colors are entered in.

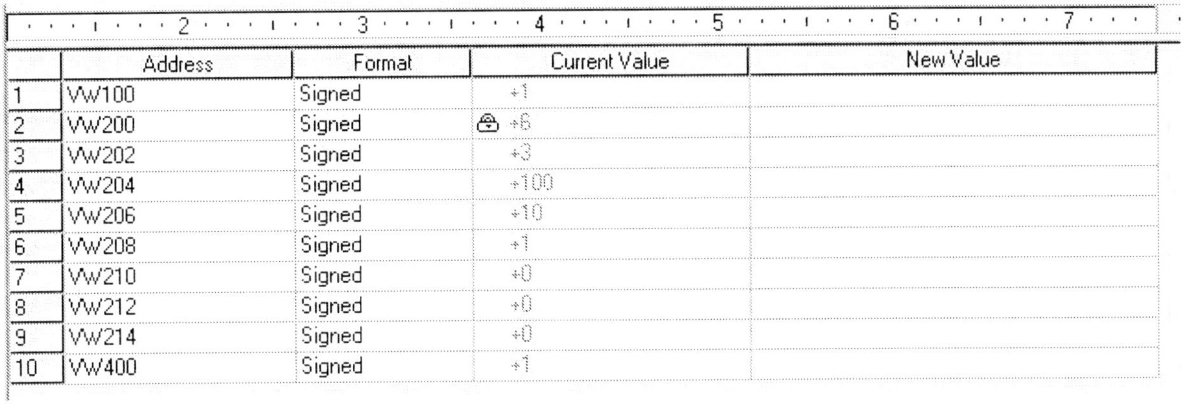

	Address	Format	Current Value	New Value
1	VW100	Signed	+1	
2	VW200	Signed	+6	
3	VW202	Signed	+3	
4	VW204	Signed	+100	
5	VW206	Signed	+10	
6	VW208	Signed	+1	
7	VW210	Signed	+0	
8	VW212	Signed	+0	
9	VW214	Signed	+0	
10	VW400	Signed	+1	

Figure 7-34 Status Chart during program operation.

Operation sequence with the ladder diagram of Figure 7-35:
1) The assembly line is stopped and the paint booth entry holding area is empty.
2) Power goes to the CPU 222.
3) The assembly line starts and sends an item to the paint booth entry holding area.
4) A bar code reader at the entrance to the holding area reads the item's bar code, interprets the code, and closes the appropriate switch connected to the CPU 222 input.
5) The CPU 222 stores the color code as the first received in a data table. The colors are stored in the CPU 222 with the following words: Red = decimal 1, Yellow = decimal 10, and Green = decimal 100.
6) In this example, suppose the entered item needs to be painted red, then VW100 will be set to 1.
7) The "AD_T_TBL" instruction places the value of VW100 in the six space long data table at VW204. The table now looks like:

Word	VW204	VW206	VW208	VW210	VW212	VW214
Value	1	0	0	0	0	0

8) The output word to VW400 is 0, since the "FIFO" instruction has not yet been told to move a value from VW204 to VW400.

9) Suppose the next entered item needs to be painted green, then the data table will be changed to:

Word	VW204	VW206	VW208	VW210	VW212	VW214
Value	1	100	0	0	0	0

10) Suppose the next entered item needs to be painted yellow, then the data table will be changed to:

Word	VW204	VW206	VW208	VW210	VW212	VW214
Value	1	100	10	0	0	0

11) Suppose the next entered item needs to be painted red, then the data table will be changed to:

Word	VW204	VW206	VW208	VW210	VW212	VW214
Value	1	100	10	1	0	0

12) The output word, VW400, has been 0 throughout this process, since the items have not operated the SPRAY BOOTH ENTRY limit switch, which would have caused the "FIFO" to unload the table.

14) When the "FIFO" operates the first time it will send the red word, 1, to VW400. Then the appropriate "Equal Integer" instruction makes the red spray paint solenoid valve receive power and go on.

15) At the same time the words will shift to the left so that the data stack will be changed to:

Word	VW204	VW206	VW208	VW210	VW212	VW214
Value	100	10	1	1	0	0

16) Note that VW210 still has a value of 1. That is not a real value. The data table now regards VW210 as having no value. To make VW210 have a meaningful value it would have to have it entered via VW100, as the other values were.

7.12.1 LADDER DIAGRAM PROGRAM

PROGRAM COMMENTS

Network 1
Network Comment: I0.0 = RED NEEDED SWITCH, MOV_W = PUTS 1, THE CODE FOR RED, IN VW100, P = LOGICALLY CONDUCTS ONCE PER I0.0 ON AND OFF

```
    I0.0              MOV_W
─────┤ ├──────┤ P ├────┤EN    ENO├────
                       │          │
                    1 ─┤IN    OUT├─ VW100
```

Network 2
Network Comment: I0.1 = YELLOW NEEDED SWITCH, MOV_W = PUTS 10, THE CODE FOR YELLOW, IN VW100, P = LOGICALLY CONDUCTS ONCE PER I0.1 ON AND OFF

```
    I0.1              MOV_W
─────┤ ├──────┤ P ├────┤EN    ENO├────
                       │          │
                   10 ─┤IN    OUT├─ VW100
```

Network 3
Network Comment: I0.2 = GREEN NEEDED SWITCH, MOV_W = PUTS 100, THE CODE FOR GREEN, IN VW100, P = LOGICALLY CONDUCTS ONCE PER I0.2 ON AND OFF

```
    I0.2              MOV_W
─────┤ ├──────┤ P ├────┤EN    ENO├────
                       │          │
                  100 ─┤IN    OUT├─ VW100
```

Network 4
Network Comment: I0.0 = RED NEEDED SWITCH, I0.1 = YELLOW NEEDED SWITCH, I0.2 = GREEN NEEDED SWITCH, P = LOGICALLY CONDUCTS ONCE PER I0.0 OR I0.1 OR I0.2 ON AND OFF, AD_T_TBL = PUTS THE VALUE OF VW100 IN THE VW200 TABLE

```
    I0.0                     AD_T_TBL
─┬───┤ ├───┬────┤ P ├────────┤EN    ENO├────
 │         │                 │            │
 │  I0.1   │          VW100 ─┤DATA        │
 ├───┤ ├───┤          VW200 ─┤TBL         │
 │         │                 │            │
 │  I0.2   │
 ├───┤ ├───┤
```

Continued

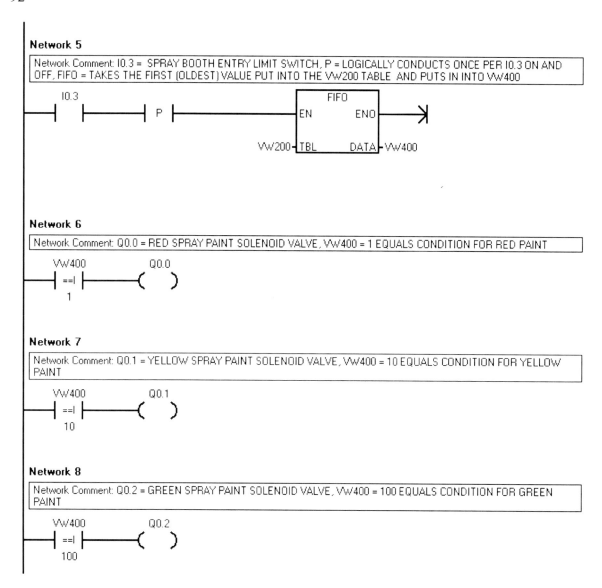

Figure 7-35 Ladder diagram for selecting spray paint colors.

7.12.2 CPU 222 CIRCUIT

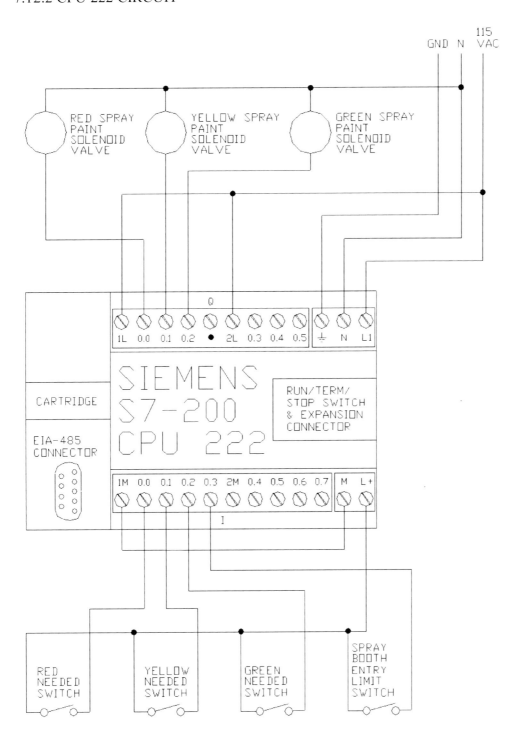

Figure 7-36 CPU 222 circuit for selecting spray paint colors.

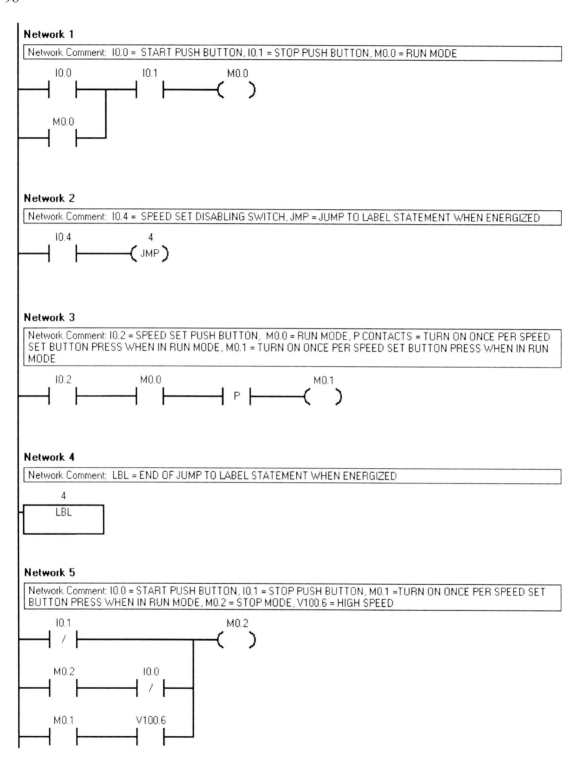

Figure 7-38 CPU 222 SBR_0 ladder diagram program, subroutine 0.

7.12.2 CPU 222 CIRCUIT

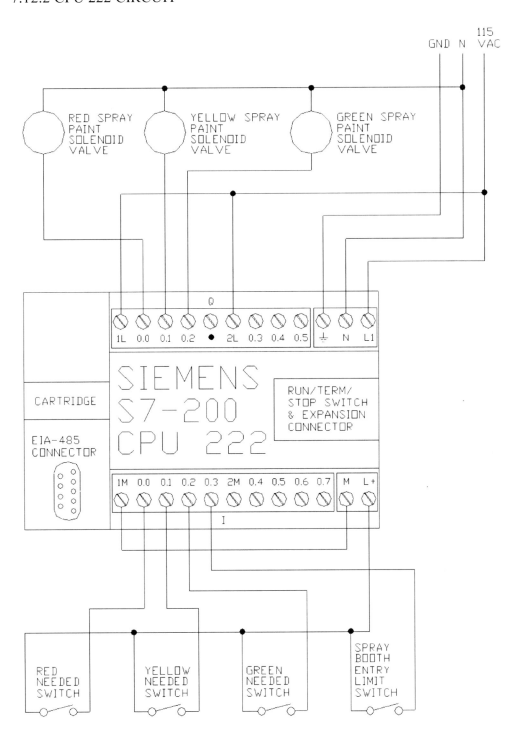

Figure 7-36 CPU 222 circuit for selecting spray paint colors.

7.13 CPU 222 LADDER DIAGRAM SUBROUTINES AND JUMP STATEMENT

In this example, the ladder diagram program of Section 7.11 is modified to demonstrate the use of subroutines and the "JMP", Jump, statement. Subroutines aid the programmer by separating ladder diagram programs into easier to understand segments and allowing program segments to be used more than once without re-typing. Usually subroutines are used with larger programs. The "JMP", Jump, statement is also demonstrated. It makes the ladder diagram program bypass the NETWORKs enclosed between it and its corresponding "LBL", Label, statement.

The "MAIN" program is in Figure 7-37. The three subroutines are in Figures 7-38, 7-39, and 7-40. They are called SBR_0, SBR_1, and SBR_2 respectively. SBR_1 and SBR_2 contain the same statements as in Figure 7-32 of Section 7.11. SBR_0 has the same statements as in Figure 7-32 but also has "JMP" and "LBL" statements.

The "JMP" and its corresponding "LBL" are both marked with the same number. They will cause the program to skip around the statement with the SPEED SET push button contact when the I0.4 switch is closed, disabling the SPEED SET push button.

To create a ladder program that uses subroutines:
1) Create space for subroutines in the ladder program by going to "Edit" in the main toolbar and select "Insert". In "Insert" select "Subroutine". This will attach a subroutine to your "MAIN" program. The subroutine will be accessible by a Tab at the bottom of the "MAIN" program window. The first subroutine will be labeled SBR_0. Repeat this two more times to create SBR_1 and SBR_2.
2) Put the statements that will call the subroutines into the "MAIN" program by using the "Call Subroutines" in the "Instruction Tree". When the spaces for the subroutines were created in 1) "CALL" statements were created in "Call Subroutines".
3) When each subroutine Tab is left clicked its subroutine editing area becomes available. Enter ladder programs into those in the same way you would into an ordinary ladder diagram program.
4) When the "MAIN" program is run, its subroutines will be called as directed.

When the program is run, be sure to put a value of 2#0000_0001 in VB100, as was done in Section 7.11.

7.13.1 LADDER DIAGRAM PROGRAM

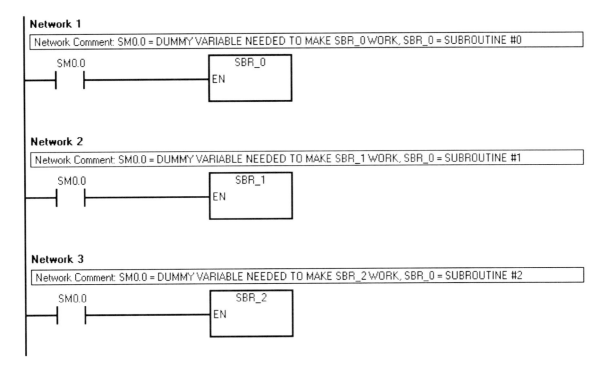

Figure 7-37 CPU 222 "MAIN" ladder diagram program.

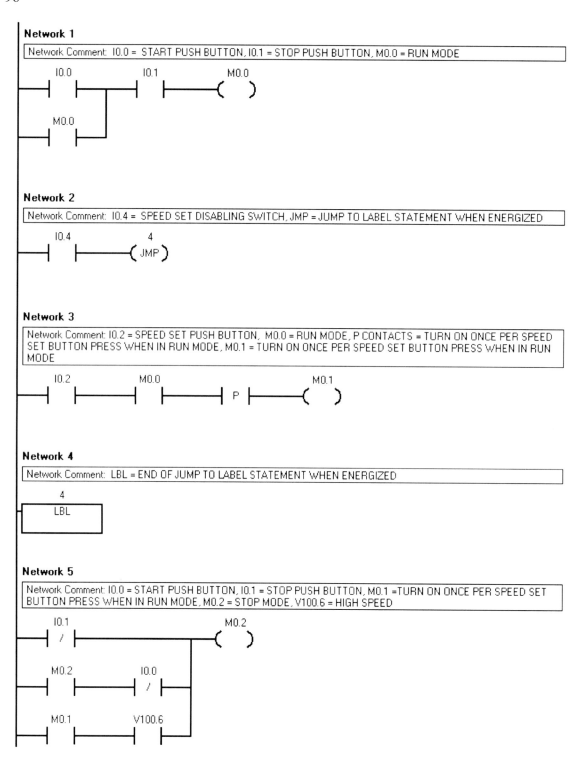

Figure 7-38 CPU 222 SBR_0 ladder diagram program, subroutine 0.

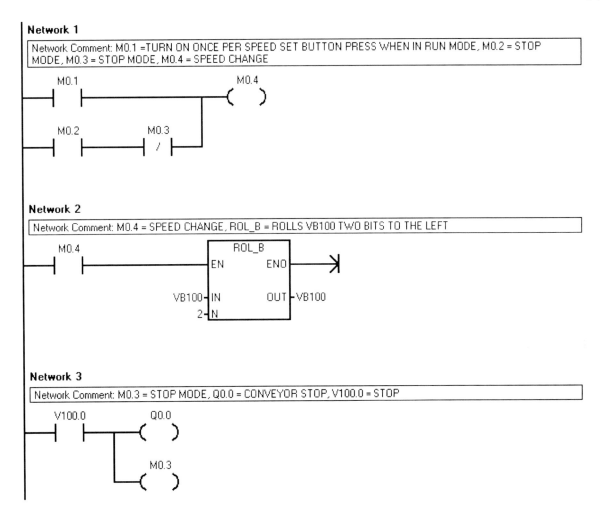

Figure 7-39 CPU 222 SBR_1 ladder diagram program, subroutine 1.

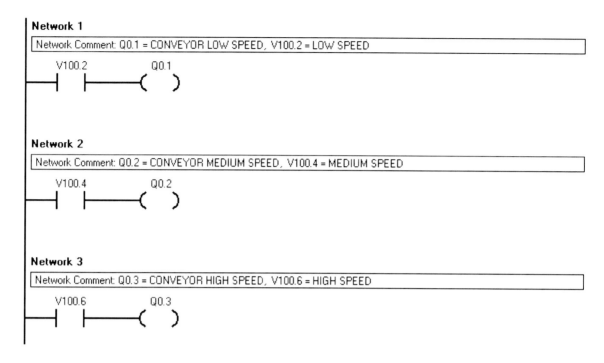

Figure 7-40 CPU 222 SBR_2 ladder diagram program, subroutine 2.

7.12.2 CPU 222 CIRCUIT

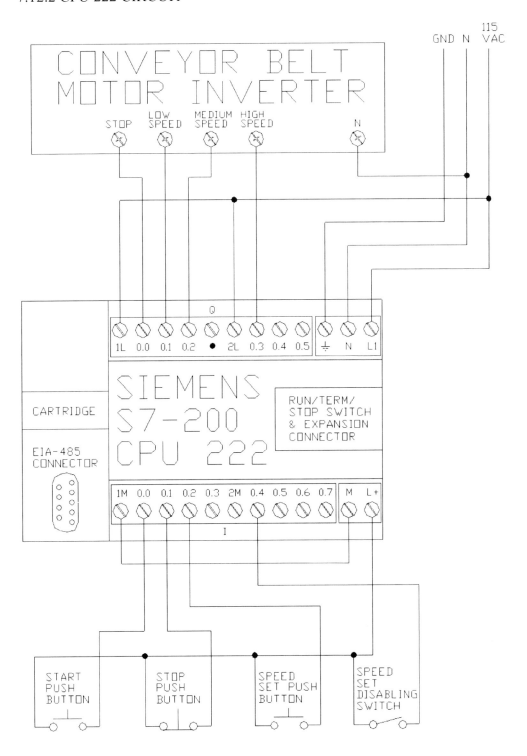

Figure 7-41 CPU 222 circuit for the conveyor belt speed controller with a speed set disabling switch.

7.14 CPU 222 LADDER AND POWER DIAGRAMS FOR A SYSTEM THAT USES HIGH-SPEED COUNTING TO MEASURE AND CONTROL MOTOR SPEED

In this example, the CPU 222 High-Speed Counter is used to measure the forward and reverse rotation increments (counts) per time produced by an induction motor driven encoder over 15-second time intervals. The number of counts is used to set the output frequency of an inverter that drives the motor.

The CPU 222 ladder diagram Count Up, "CTU", and Count Down, "CTD", counters can only count one count per program scan. High-Speed Counters count independently of the ladder diagram program, allowing them to count much faster. The CPU 222 High-Speed Counter can handle counting input frequencies of up to 20 kHz with two-phase counting. With single-phase counting the maximum input frequency is 30 kHz. A two-phase counter can determine encoder rotation direction. Rotation direction is either clockwise or counterclockwise as seen from the encoder shaft end. A single-phase counter can not determine rotation direction. The two-phase configuration produces four counts per cycle. With the two-phase configuration and an encoder producing 100 counts per revolution, motor speeds could be measured up to (20000/4/100)x60 = 3000 rpm.

The High-Speed Counter can not be accessed directly in a ladder diagram. It is accessed by a special interrupt routine that a "MAIN" ladder diagram program or a subroutine calls. An interrupt routine is similar to a subroutine, however an interrupt routine operates much faster. Once started, the High-Speed Counter operates continuously without regard to the ladder program's scan or scan rate.

The new material demonstrated here is:
1) The "HSC <High-Speed Counter> Wizard" for setting up the High-Speed Counter.
2) The "HSC", High-Speed Counter, instruction.
3) The "⊣ <=D ⊢", Less Than or Equal Double Integer, instruction.
4) The "⊣ >=D ⊢", Greater Than or Equal Double Integer instruction.

Procedure for writing the ladder diagram program:
1) Draw the ladder diagram "MAIN" program as shown in Figure 7-42, but leave out the "CALL" for the "HSC_INIT" High-Speed Counter interrupt routine. Save the program.
2) Start the "High-Speed Counter Wizard". This can be found in the "Instruction Tree" window under "Wizards".
3) Select "HC0" as the counter to configure.
4) Select "Mode 10". This is for an A/B quadrature counter with a Reset input, but no Start input.
5) Leave the rest of the selections as they are.

6) "Finish" the wizard and it will automatically create an interrupt routine named "HSC_INIT". This routine can be accessed by a tab from the "Program Block" window. It can be seen in Figure 7-43.

7) In the "Instruction Tree" go to the "Call Subroutines" and open it. Drag and place the call for "HSC_INIT" into the ladder diagram of Figure 7-42.

8) Save and run the program as usual.

9) Details on the High-Speed Counter and its Wizard can be seen in the Help files of Step 7-Micro/WIN and in the *S7-200 Programmable Controller Systems Manual*.

Operation sequence with the ladder diagram of Figure 7-42:

1) Power goes to the CPU 222.

2) The High-Speed Counter starts counting and the "T37" timer starts timing.

3) After 15 seconds the "⊣<=D⊢" and "⊣>=D⊢" instructions compare the count to two ranges. If the count is from 0 to 7 then INVERTER PRESET SPEED #1 will be enabled. If the count is greater than or equal to 8 then INVERTER SPEED #2 will be enabled. The count limits here are low so that they can be met with the manually operated switches of the "teaching setup".

4) The CPU 222 will determine the direction of rotation by the order of the turning on and off of the inputs I0.0 and I0.1. If the order of turning on and off is I0.0 on, I0.1 on, I0.0 off, I0.1 off, I0.0 on, I0.1 on and so on, the count number will be positive. If the order of turning on and off is I0.1 on, I0.0 on, I0.1 off, I0.0 off, I0.1 on, I0.0 on and so on, the count number will be negative. This can be verified with the "teaching setup".

5) After the ladder diagram program turns on the appropriate preset inverter speed (inverter output frequency), the program resets the timer to zero and restarts the high-speed counter from zero.

6) Counting is again done independently of the ladder diagram program.

7) At any time the High-Speed Counter can be reset to zero by turning on I0.2.

7.14.1 LADDER DIAGRAM PROGRAM

Network 1

Network Comment: HSC_INIT = CALL OF THE HIGH-SPEED COUNTER INTERRUPT ROUTINE (HSC_INIT), SM0.1 CONTACTS = LOGICALLY CONNECT ONLY FOR THE FIRST SCAN, T37 CONTACTS = CLOSE EVERY 15 SECONDS TO RESET THE HIGH-SPEED COUNTER

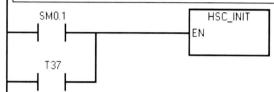

Network 2

Network Comment: T37 CONTACTS = CLOSE EVERY 15 SECONDS TO RESET THE T37 TIMER, T37 R = RESETS TIMER WHEN T37 CONTACTS LOGICALLY CONDUCT

```
   T37         T37
───┤ ├────────( R )
                1
```

Network 3

Network Comment: SM0.0 = ALWAYS ON DUMMY INSTRUCTION NEEDED TO MAKE TIMER T37 WORK, T37 TIMER = TIMER THAT TURNS ON AFTER 15 SECONDS

```
   SM0.0              T37
───┤ ├──────────┤IN       TON├
                │             │
           150──┤PT    100 ms │
```

Network 4

Network Comment: MOV_DW = MOVES THE HIGH SPEED COUNTER COUNT, HC0, TO DOUBLE WORD (32 BITS) VD100, T37 CONTACTS = CLOSE EVERY 15 SECONDS TO OPERATE MOV_DW

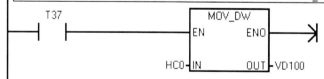

Continued

Network 5

Network Comment: Q0.0 R = RESETS OUTPUT Q0.0, PRESET SPEED #1, TO OFF, Q0.0 S = SETS AND LATCHES OUTPUT Q0.0, PRESET SPEED #1, TO ON, Q0.1 R = RESETS OUTPUT Q0.1, PRESET SPEED #2, TO OFF, Q0.1 S = SETS AND LATCHES OUTPUT Q0.1, PRESET SPEED #2, TO ON, T37 CONTACTS = CLOSE EVERY 15 SECONDS TO OPERATE COMPARISON CIRCUITS, >= D 0 CONTACTS = LOGICALLY CONDUCTS WHEN VD100 IS GREATER THAN OR EQUAL TO 0, >= D 7 CONTACTS = LOGICALLY CONDUCTS WHEN VD100 IS GREATER THAN OR EQUAL TO 7, <= D 6 CONTACTS = LOGICALLY CONDUCTS WHEN VD100 IS LESS THAN OR EQUAL TO 6

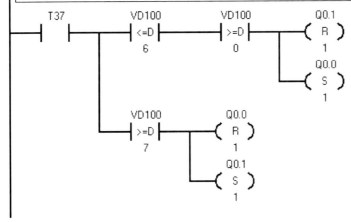

Figure 7-42 Ladder diagram program for the encoder control circuit.

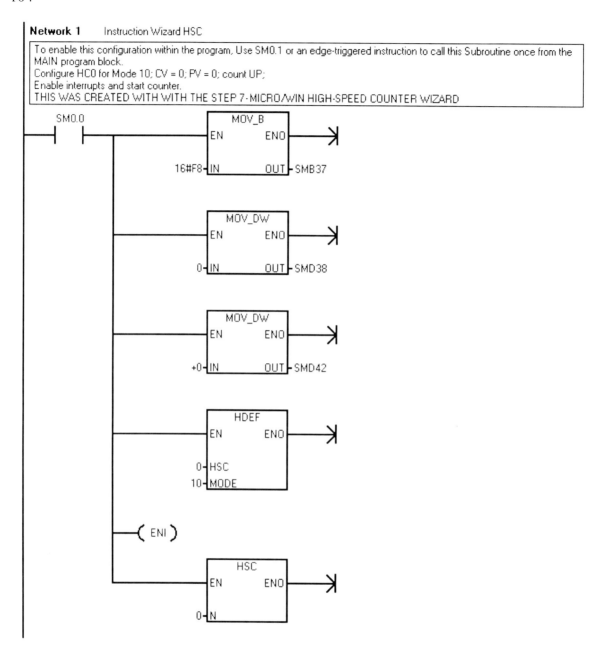

Figure 7-43 "HSC_INIT" High-Speed Counter interrupt routine that was created by the "HSC Wizard".

7.14.2 CPU 222 CIRCUIT

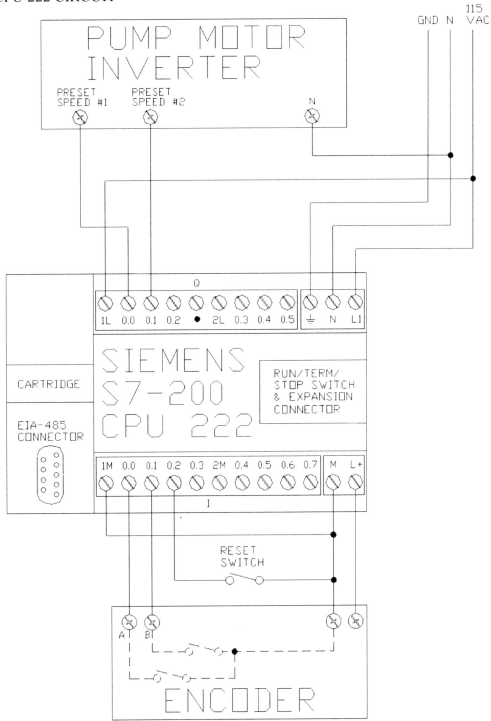

Figure 7-44 CPU 222 circuit with an encoder.

7.14.3 POWER CIRCUIT

The same power circuit is used here as was used in Section 7.9. See Figure 7-26, page 75.

8.0 OTHER SOURCES OF INFORMATION

8.1 USEFUL REFERENCE BOOKS

Cox, R. A., 2006. *Technician's Guide to Programmable Controllers,* 5th Ed. Albany, NY: Delmar Thomson Learning, price $69.95.
This is an excellent book. Unfortunately, its illustrative examples are only of Allen-Bradley and Modicon PLCs. However, even with the book's lack of Siemens PLCs, it is a worthwhile purchase.

Siemens AG, 2004. *Products for Totally Integrated Automation and Micro Automation Catalog ST 70-2005*, Nürnberg, Germany, Order No. E86060-K4670-A111-A9-7600, free from distributors or on the Internet at
http://www1.siemens.cz/ad/current/layers/data/downloads/c003as/as_katalogy/as_katalogy_ST70/ST70_en.pdf
This is a 640 page catalog of Siemens Simatic PLC hardware and software. It would help the reader better understand and select from the wide range of Siemens control equipment.

Siemens AG, 2004. *Simatic S7-200 Programmable Controller System Manual*, Nürnberg, Germany, Order No. 6ES7298-8FA24-8BH0. A paper copy and a copy on CD are included with the Step 7-Micro/WIN CD that comes with a CPU 222 starter kit. An older version of it is available for free download from the Internet at
http://www.ad.siemens.com.cn/download_test/manual/en/as/manual/S7200N_e.pdf
This manual is 515 pages long. It is good for reference, but is not written for beginners.

Siemens AG, *Basics of PLCs*, Alpharetta, GA, Order No. STTM-EP10F-0605. A free paper copy may be available from your Siemens distributor. A free online version is available at
http://www.sea.siemens.com/step/templates/lesson.mason?plcs:1:1:1
This is an 88 page tutorial designed for beginners.

8.2 SIEMENS CONTACTS

Siemens Aktiengesellschaft
Automation and Drives
Industrial Automation Systems
Postfach 4848
90327 Nürnberg
GERMANY
www.siemens.com/automation

Siemens Energy & Automation, Inc.
3333 Old Milton Parkway
Alpharetta, GA 30305
U.S.A.
1-800-964-4114

Siemens offers its own PLC classroom courses and web-based training. Look on the web at http://www.sitrain.siemens.com/f_0.html. The classroom courses are very good, but may be too expensive for some individuals. Most of the students in the classroom courses have their tuition paid by their employers. The web-based training costs less and might be more in the range of an individual's budget.

Siemens Energy & Automation provides up to one hour of free technical phone support for their equipment and software. Call 800-333-7421. They will also answer technical questions by email. There is no stated limit on the email support. The email address to send to is: techsupport.sea@siemens.com.

Local Siemens distributors and representatives will sometimes provide free help to those using or teaching themselves how to use Siemens PLCs. The Siemens corporate office can let you know which distributors and representatives serve your area.

8.3 ELECTRICAL DISTRIBUTORS THAT SELL SIEMENS PLCS

It is best to purchase a new CPU 222 from a distributor rather than buy a used one. If you purchase a new CPU 222 from a distributor it would be easier to call them for technical help and Siemens will also provide free help after the sale. Also, the distributors sometimes offer free instructional/sales PLC seminars.

9.0 APPENDIX

9.1 RELAY LADDER AND POWER DIAGRAM SYMBOLS

Relay or contactor coil

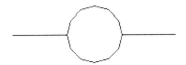

Normally open relay or contactor contacts

Normally closed relay or contactor contacts

Normally open push button

Normally closed push button

Normally open single pole switch or disconnect

Normally closed single pole switch

Circuit breaker

Pilot light

Three-phase induction motor

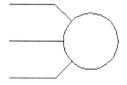

9.2 "STEP 7-MICRO/WIN" LADDER DIAGRAM PROGRAM MEMORY AREAS

1) I = Process image input register. Usually used for entering 1's and 0's into the PLC by electrical connection points. The CPU 222 base unit has eight physical connection points, I0.0 to I0.7. "I" can also be used with bytes (8 bits, i.e. IB1), words (16 bits, i.e. IW1), and double words (32 bits, i.e. ID1). If expressed in bits, the CPU 222 has 157 of these. They are I0.0 to I15.7.

2) Q = Process image output register. Usually used for turning on and off controlled devices via electrical connection points on the PLC. The CPU 222 base unit has six physical connection points, Q0.0 to Q0.5. "Q" can also be used with bytes (8 bits, i.e. QB1), words (16 bits, i.e. QW1), and double words (32 bits, i.e. QD1). If expressed in bits, the CPU 222 has 157 of these. They are Q0.0 to Q15.7.

3) AIW = Analog inputs (read only). These are 16-bit words that the CPU 222 could receive from an analog input module. There are 16 of them. They range from AIW0, AIW2,…to AIW30.

4) AQW = Analog outputs (write only). These are 16-bit words that the CPU 222 could send to an analog output module. There are 16 of them. They range from AQW0, AQW2,…to AQW30.

5) V = Variable memory. They can be used with bits, 8-bit bytes, 16-bit words, or 32-bit double words. They are often used with counters, timers, and math calculations. If they are expressed in 8-bit bytes (VB__) the CPU 222 has 2048 of these, ranging from VB0 to VB2047.

6) L = Local memory. This is similar to "V", variable memory. However local memory applies to specific portions of ladder programs. For example, a local memory word in one subroutine can not be accessed in another subroutine. If they are expressed in 8-bit bytes (LB__) the CPU 222 has 64 of these, ranging from LB0 to LB63 available to each program portion.

7) M = Bit memory. They are for use inside of ladder diagram programs. They can be used with bits, 8-bit bytes, 16-bit words, or 32-bit double words. If expressed in bits, the CPU 222 has 318 of these. They are M0.0 to M31.7.

8) SM = Special memory (read only). These are for communicating between the CPU and the program. They can be accessed as bits, 8-bit bytes, 16-bit words, or 32-bit double words.

9) T = Timers, retentive on-delay and on/off delay. Timer functions measure elapsed times and then store it as 16-bit words. They also produce single bit values indicating whether a timer has completed timing. There are 256 available timers with the CPU 222.

10) C = Counters. Counter functions count and store values as 16-bit words. They also produce single bit values indicating whether a counter has completed counting. There are 256 available counters with the CPU 222.

11) HC = High-speed counters. High-speed counters count and store values in 32-bit double words. They operate independently of the CPU 222's program. There are a maximum of six high-speed counters in a CPU 222. Count values are stored in HC0 to HC5.

12) AC = Accumulator registers. These are read/write memory devices. They can store up to 32-bit double words. There are four of them available in the CPU 222, AC0, AC1, AC2, and AC3.

9.3 "STEP 7-MICRO/WIN" LADDER DIAGRAM PROGRAM INSTRUCTION SYMBOLS

These commonly used instruction symbols have been taken from the examples of Chapter 7.0. Their operation is described for the use they had in the examples. The instructions have more capabilities than seen in chapter 7.0. For example, here an instruction may be seen using only I, Q, or M variables, but probably it is capable of operating with many other variables. More information on each instruction can be seen in "Step 7-Micro/WIN" by left clicking on it in the "Instruction Tree" and pressing F1.

.........BIT LOGIC.........

Normally Open Instruction:
This is similar to a relay diagram's normally open relay contact. When its variable is "on" (bit = 1) it will have logical continuity through it. When its variable is "off" (bit = 0) it will not have logical continuity. In the figure below, I0.0 is the name of the variable that controls it. In this book's examples, C, M, Q, SM, T, and V variables also control the **Normally Open** instruction. It can be seen in the ladder diagram program of each example.

```
     I0.0
───┤   ├───
```

Normally Closed Instruction:
This is similar to a relay diagram's normally closed relay contact. When its variable is "off" (bit = 0) it will have logical continuity through it. When its variable is "on" (bit = 1) it will not have logical continuity. In the figure below, Q0.1 is the name of the variable that turns it on and off. In this book's examples, I, M, and T variables also control the **Normally Closed** instruction. It can be seen in the ladder diagram programs of Figures 7-4, 7-8, 7-11, 7-13, 7-18, 7-20, 7-32, 7-38, and 7-39.

```
     Q0.1
───┤ / ├───
```

Positive Transition Instruction
This instruction does not have an equivalent in an ordinary relay ladder connection diagram, although it is a similar to a relay's contacts. When it receives an "on" logic signal from the left, it logically conducts for one ladder diagram program scan. For it to logically conduct during another scan it must have the logic signal from its left turned off and on again. It can be seen in Figures 7-29, 7-32, 7-35, and 7-38.

```
───┤ P ├───
```

Output Instruction:

This is similar to a relay diagram's relay coil. When it receives an "on" logic signal from the left, it turns on its assigned internal or external output. In the figure below, Q0.0 is the name of the variable it turns on and off. In this book's examples, the **Output** instruction also controls M variables. **Output** instructions can be seen in the ladder diagram programs of each example.

```
    Q0.0
───(   )
```

Set Instruction:

This instruction does not have an equivalent in ordinary relay ladder connection diagrams. It is turned on by an "on" logic signal, like the **Output** instruction. However, in an operating program, once it is on, it remains on (set or latched on) until it stops receiving an "on" logic signal and its corresponding **Reset** instruction receives an "on" logic signal. In the figure below, M0.0 is the name of the variable it turns on. The 1 under the "S" shows that only 1 bit is turned on. Higher numbers would show that more bits would be turned on, starting at M0.0. It can be seen in Figures 7-8, 7-18, 7-22, 7-24, and 7-42.

```
    M0.0
───( S )
     1
```

ReSet Instruction:

This instruction does not have an equivalent in ordinary relay ladder connection diagrams. It is turned on by an "on" logic signal, like the **Output** instruction. When it is turned on it causes its corresponding **Set** instruction to turn off, provided the **Set** instruction is no longer receiving an "on" logic signal. It also may be used to reset counters and timers. In the first figure below, M0.0 is the name of the **Set** variable that will be reset. In the second figure below, T5 is the name of the timer whose time will be reset. The 1 under the "R" shows that only 1 bit is reset. Higher numbers would show that more bits would be turned on, starting at M0.0. It can be seen resetting the **Set** instruction in Figures 7-8, 7-18, 7-22, 7-24, and 7-42. It can be seen resetting timers in Figures 7-20 and 7-24.

```
    M0.0                T5
───( R )            ───( R )
     1                   1
```

........COMPARE........

Compare instructions do not have equivalents in ordinary relay ladder connection diagrams. They logically conduct when the condition of the compare instruction is met. In this book, compared values are integers, bytes, words, or double words. Step 7-Micro/WIN can compare many other quantities.

Less Than Integer Instruction
    Logically conducts when VW100 (a 16-bit word) is less than 6. It can be seen in Figures 7-15 and 7-22.

```
    VW100
───┤ <I ├───
      6
```

Greater Than or Equal Integer Instruction
    Logically conducts when C1 (a decimal integer) is greater or equal to 4. It can be seen in Figure 7-22.

```
     C1
───┤ >=I ├───
      4
```

Greater Than Integer, Instruction:
    Logically conducts when VW100 (a 16-bit word) is greater than 2. It can be seen in Figure 7-24.

```
    VW100
───┤ >I ├───
      2
```

Less Than or Equal Integer Instruction:
    Logically conducts when VW100 (a 16-bit word) is less than or equal to 4. It can be seen in Figure 7-24.

```
    VW100
───┤ <=I ├───
      4
```

Equal Integer Instruction:
    Logically conducts when VW400 (a 16-bit word) is equal to 1. It can be seen in Figure 7-35.

```
    VW400
───┤ ==I ├───
      1
```

Less Than or Equal Double Integer Instruction:
Logically conducts when VD100 (a 32-bit double word) is less than or equal to 6. It can be seen in Figure 7-42.

```
   VD100
──┤ <=D ├──
     6
```

Greater Than or Equal Double Integer Instruction:
Logically conducts when VD100 (a 32-bit double word) is greater than or equal to 0. It can be seen in Figure 7-42.

```
   VD100
──┤ >=D ├──
     0
```

.........COUNTERS.........

CTU, Count Up Instruction:
This instruction does not have an equivalent in an ordinary relay ladder connection diagram. When it receives an "on" and "off" logic signal to its upper left input, it increases its stored count value by one. The count value is stored as a 16-bit word in the C# above the count instruction. When the "CTU" has counted to its "PV" (Preset Value) it supplies an "on" logic signal to contacts that have the same C# notation. Once on, the C# logic signal stays on until an "on" logic signal is supplied to the counter's "R" (Reset). The "on" logic signal to the "R" input also resets the counter to 0. Stored counts are retentive through power on/off cycles. In the figure below, the count value is stored in C1. The "PV" value shows that the counter will count to 6 before the C1 will supply an "on" logic signal to contacts with C1 above them. "CTU"s can be seen in Figures 7-11, 7-13, 7-15, 7-20, and 7-24.

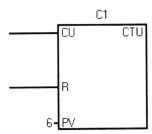

HDEF, High-Speed Counter Definition Instruction:

This specifies the high-speed counter's type and mode. In present S7-200s **HDEF** and the interrupt routine it goes in are configured by the **High-Speed Wizard**. **HDEF** can be seen in Figure 7-43.

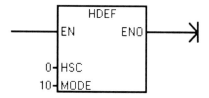

HSC, High-Speed Counter Instruction:

This configures and controls the high-speed counter. The number next to N specifies the high-speed counter number. The modes of operation are set with the **HDEF** instruction and the high-speed counter interrupt routine. High-speed counters count independently of the ladder program at much higher speeds than ordinary ladder program counters. The output count of an **HSC** is HC#. For the example below, the count number would be HC0, where HC0 is a 32-bit double word. It can be seen in Figure 7-43.

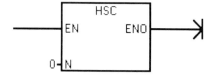

.........INTEGER MATH.........

ADD_I, Add Integer Instruction:

This instruction does not have an equivalent in an ordinary relay ladder connection diagram. When its upper left input receives an "on" logic signal it adds the integer value at "IN1" to the value at "IN2" and puts the sum into the variable next to the "OUT". In the figure below, the values being added are the count value, C1, and 3. The sum is stored in VW100. It can be seen in Figure 7-24.

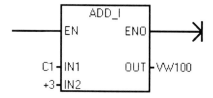

SUB_I, Subtract Integer Instruction:

This instruction does not have an equivalent in an ordinary relay ladder connection diagram. When its upper left input receives an "on" logic signal, it subtracts the integer value at "IN1" by the value at "IN2" and puts the sum in the variable next to the "OUT". In the figure below, the count value of C1 is subtracted by the count value of C2 and the difference put in VW100. It can be seen in Figure 7-15.

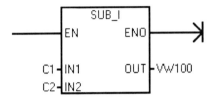

INC_B, Increment Byte Instruction:

This instruction does not have an equivalent in an ordinary relay ladder connection diagram. When its upper left input receives an "on" logic signal, it increments the value at "IN" by 1 and writes the result into the variable at "OUT". In the figure below, the count value of VB100 is incremented by 1 each time its upper left input receives an "on" logic signal. It can be seen in Figure 7-29.

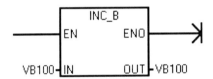

.........INTERRUPT.........

Enable Interrupt Instruction:

This instruction enables all attached interrupt events. In the example in Figure 7-43, it enables the high-speed counter.

—(ENI)

……….MOVE……….

MOV_B, Move Byte Instruction:

This instruction does not have an equivalent in an ordinary relay ladder connection diagram. When it receives an "on" logic signal to its upper left input it moves the 8-bit byte value stored at "IN" to the variable at "OUT". In the figure below, the integer value being moved is 6 and it goes across to variable AC1. It can be seen in Figures 7-13 and 7-43.

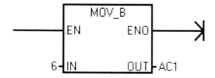

MOV_W, Move Word Instruction:

This is the same as the MOV_B instruction, except that it moves 16-bit words rather than 8-bit bytes. It can be seen in Figure 7-15.

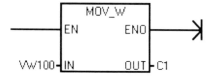

MOV_DW, Move Double Word Instruction:

This is the same as the MOV_B instruction, except that it moves 32-bit double words rather than 8-bit bytes. It can be seen in Figures 7-42 and 7-43.

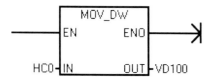

.........PROGRAM CONTROL.........

Jump and Label Instruction:

These instructions do not have equivalents in an ordinary relay ladder connection diagram. When the "JMP" instruction receives an "on" logic signal to its left input, it will make the ladder diagram program ignore the rungs between it and its corresponding "LBL" instruction. In the figures below, 4 is the designation of the corresponding "JMP" and "LBL" instructions. "JMP" and "LBL" can be seen in Figure 7-38.

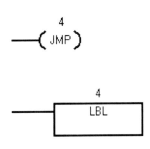

Stop Instruction:

This instruction stops a program by transitioning a S7-200 program from "RUN" to "STOP". It can be seen in Figure 7-18.

.........SHIFT/ROTATE.........

SHL_B, Shift Left Byte Instruction:

This instruction does not have an equivalent in an ordinary relay ladder connection diagram. When it receives an "on" logic signal to its upper left input it slides bits in the byte variable next to "IN" to the left. When the bits exceed the left limit, they are lost. In the figure below, bits of VB100 are shifted one to the left (as directed by the 1 next to "N"). The new value of VB100 appears next to "OUT". The spaces emptied as bits are shifted are filled with zeros. It can be seen in Figure 7-29.

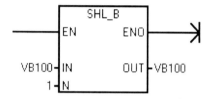

ROL_B, Rotate Left Byte Instruction:

This instruction does not have an equivalent in an ordinary relay ladder connection diagram. When it receives an "on" logic signal to its upper left input, it rotates bits of the variable next to "IN" to the left, and then, after the bits reach the left extreme, back to the right. The bits are rotated the number of bits written next to "N". After rotating, the value is stored in the variable next to the "OUT". In this case, the variable VB100 has its bits rotated two bits (as directed by the 2 next to the "N") every time an "on" logic signal goes to the upper left input. It can be seen in Figures 7-32 and 7-39.

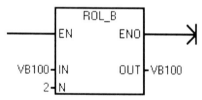

.........TABLE.........

FIFO, First In First Out Instruction:
    This does not have an equivalent in an ordinary relay ladder diagram. When it receives an "on" logic signal to its upper left input, the **FIFO** instruction moves the oldest entry in a data table to output memory, and shifts the other entries up one location. The first word of the data table is written next to "TBL". The oldest entry is moved to the variable next to "DATA". **FIFO** is used in programs with the, "AD_T_TBL", Add to Table instruction. **FIFO** can be seen in Figure 7-35.

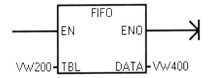

AD_T_TBL, Add to Table Instruction:
    This does not have an equivalent in an ordinary relay ladder diagram. When it receives an "on" logic signal to its upper left input, the "AD_T_TBL" instruction adds the data of the variable next to "DATA" to a table that starts at the variable next to "TBL". Data is added to the table each time that there is an "on" logic signal to the upper left input of "AD_T_TBL". The variable next to "DATA" (VW200 in the figure below) stores a number that defines the number of words stored in the table. The value for the variable (VW200 in the figure below) can be forced into it outside of the ladder program. It is shown in Figure 7-35 and explained in more detail in Section 7.12.

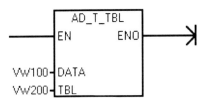

.........TIMERS.........

TON, On-Delay Timer Instruction:

This instruction is similar to a timer relay in an ordinary relay ladder connection diagram. When its upper left input receives an "on" logic signal, it starts timing. It stops timing and clears its time when it no longer receives an "on" logic signal. When timing, if the instruction reaches the time next to "PT" (Preset Time), it turns on contacts that have the same T# above them as above the **TON**. Resetting the timer is done with a "ReSet" instruction. The timer's time base is in the lower right corner of the **TON**. This time base is multiplied by the number put next to the PT to determine the time when the timer will turn on. In the figure below, the timer will turn on the contacts marked T37 when it has timed to 30 x 100 ms = 3 seconds. It can be seen in Figures 7-8, 7-18, 7-22, 7-24, and 7-42.

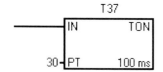

TONR, Retentive On-Delay Timer Instruction:

This instruction is the same as the **TON** except that its time is retentive. The timer will remember the time even if power is lost to its PLC or it is not receiving an "on" logic signal. When it again receives an "on" logic signal, it will start timing from that time. It can be seen in Figure 7-20.

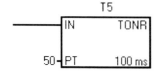

.........CALL SUBROUTINES AND INTERRUPT ROUTINES.........

These instructions do not have equivalents in ordinary relay ladder connection diagrams. When a subroutine instruction or an interrupt routine instruction receives an "on" logic signal at its left input, the ladder program directs the PLC to that subroutine or interrupt routine. After the actions called for in the subroutine are completed, the PLC is directed back to the main ladder program. After actions called for in the interrupt routine are started or carried out, the PLC is directed back to the main ladder diagram program. A call for a subroutine can be seen in Figure 7-37. A call for an interrupt routine can be seen in Figure 7-42.

9.4 PROGRAMMABLE CONTROLLER GLOSSARY

Absolute Address: Address that uses the memory area and bit or byte location to identify the address. I0.0 is an example of an absolute address. I0.0 refers to the input connection point on the base unit of a S7-200. Absolute Addresses can be assigned to combinations of alphanumeric addresses called Symbolic Addresses.

Address: Identifying numbers and/or letters that designate a particular I/O location on a PLC, designate a particular device controlled by a PLC, or designate a memory location in a PLC.

Adapter Board: Module that is used on external devices to allow them to be connected to a PLC's data bus.

ASCII (American Standard Code for Information Interchange): USA Standard Code for Information Exchange. It is a 7-bit character code based on the English alphabet. It defines 128 standard codes. 95 of the codes are printable characters and the other 33 are control codes. ASCII is one of the most successful software standards. Since most computers use at least 8-bit bytes, most computers now use an "Extended ASCII" version. "Extended ASCII" uses eight bits to define 256 codes. The "Extended ASCII" versions include "ASCII" within them.

AS-I (AS-Interface or Actuator-Sensor Interface): This is the simplest of the industrial networking protocols. It is designed for connecting on/off devices like actuators and sensors on to a single cable. It can be used with Profibus and the Industrial Ethernet.

Bit: Generally speaking, this is a binary digit, a 1 or 0. With PLCs using a ladder diagram programs, this is considered an "on" or "off". An "on" is a continuous path across a ladder network (rung). An "off" is a break in the circuit of a ladder network (rung).

Byte: Unit of binary data storage. There are 8 bits in a byte.

Cascading: Programming technique of feeding a timer output into a counter input to extend the timing range of the PLC beyond that of the timer alone.

CFC (Continuous Function Chart): High level programming language used on S7-300 and S7-400 PLCs.

CP (Communications Processor): Module or card to connect a PC or PLC to a communications bus.

CPU: 1) (Central Processing Unit) Major component of a computer system with the circuitry to control the interpretation and execution of instructions. Every PLC contains a Central Processing Unit. 2) Without regard for the usual definition of CPU, Siemens uses CPU as a prefix to differentiate its S7-200 PLCs, the CPU 221, CPU 222, etc. The different S7-200s may have the same Central Processing Unit electronics.

Contactor: Like an electromagnetic relay, but switches larger voltages and currents. Sizes range from that of a book to that of an automobile. Smaller contactors are used to handle motor or lighting loads. The largest contactors are in power company switchyards and can handle thousands of volts and amps. In schematics, contactors are represented with the same symbols as relays.

Double Word: Unit of binary data storage that is twice the size of a word. With Step 7-Micro/WIN double words use 32 Bits.

EEPROM: Electrically erasable programmable read-only memory.

EIA-485 (Electronic Industries Association-485, formerly RS-485): See Section 3.1.

Enabled: Term to indicate that an instruction has been activated.

Ethernet: Family of computer network technologies for Local Area Networks (LANS). It defines physical signaling standards, like a star-topology of twisted wire pairs, and a common addressing format. Ethernet is standardized by IEEE 802.3.

FBD (Function Block Diagram): PLC programming language that consists of connected function blocks. The blocks are connected together by lines. Each function block acts on the data input to it and outputs data to another function block or to a PLC output. The IEC standard for FBDs is 61131.3.

File: Group of consecutive words.

Firmware: Programmable controller instructions that are embedded in the hardware, stored in PROM, EPROM or EEPROM devices, and are generally not modifiable by the user.

Flash Memory: Variation of EEPROM, electrically erasable programmable read-only memory.

Freeport Mode: Mode that the S7-200 can be switched to by program statements and by turning the S7-200 switch to the RUN position. In the Freeport mode, the S7-200's integrated 9 socket D sub EIA-485 connector can be used to communicate with many types of intelligent devices. It can implement user-defined communications protocols. Freeport mode supports ASCII and binary protocols.

FTP (File Transfer Protocol): Protocol for transferring files on the internet.

Function Block Diagram: See FBD

IEC 1131: Old number for the current IEC 61131.

IEC 61131: Standard for programmable logic controllers. The major PLC manufacturers all attempt to bring their PLCs in compliance with it.

IEC 61131-3: Section of the IEC 61131 that deals with programmable controller languages. It specifies syntax, semantics, and display. There are five languages specified. They are:
- Ladder Diagram (LAD)
- Function Block Diagram (FBD)
- Instruction List (IL)
- Structured Text (ST)
- Sequential Function Charts (SFC)

IL (Instruction List): Also called STL, Statement List. This is a text based programming language.

Industrial Ethernet: Name given to the Ethernet protocol used in an industrial environment for automation. See Section 3.2.

Instruction List Programming: See IL.

Instruction Set or Instructions: Program statements used to manipulate data received by the PLC. Some of the instructions available on PLCs are: relay-type ("on" and "off"), timer, counter, comparison, arithmetic, Boolean logic, and program control.

Internet: Worldwide network of interconnected computer networks that transmit data by packet switching using a standard data-oriented protocol.

Interposing Relay: Relay placed between a relay or controller output and a higher current or voltage load. For example, a PLC output with a 1.0 A maximum capability might need to control a contactor that requires a 5.0 A coil input. To do this, the PLC would control an Interposing Relay whose coil uses less than 1.0 A, but whose output contacts can handle the 5.0 A.

IP: 1) (Internet Protocol) Protocol used with the Internet. 2) (Industrial Protocol) Protocol used with the Industrial Ethernet. Note these two protocols are not the same.

I/O: Abbreviation for "Input/Output". A PLC receives data through its input terminals and sends out signals and controlling voltages through its output terminals.

Jog: State of momentarily being on or in motion. A jog push button allows a machine operator to "inch" a machine in forward or reverse as long as the jog push button is pushed. As soon as the jog push button is released the machine stops.

LAD (Ladder Diagram): Standard method for drawing relay or logic control circuits. The drawings resemble a ladder. Most PLC manufacturers use software created ladder diagrams as one of their programming languages. The IEC standard for LADs is 61131.3.

MB (Megabyte): 1,048,576 bytes of computer or PLC memory.

Modbus: Serial communications protocol developed originally by the Modicon company for use with its PLCs. It has become an open (published) system that is no longer owned by the Modicon company. See Section 3.1.3.

Module: Interchangeable plug-in electronic item, often a box, printed circuit board, or card.

MPI (Multi-Point Interface) Protocol: A Siemens Automation programming protocol.

Network: 1) Combination of electrical elements. 2) "Step 7-Micro/WIN" uses "Network" to describe ladder diagram program rungs. Each rung is labeled "Network" followed by a number.

Open Architecture: Computer or PLC design for which detailed specifications are published by the manufacturer, allowing others to produce compatible hardware and software.

PC: Personal Computer.

PG: (Programmer): Keyboard device used to input programs and data and operate a PLC. It can be mounted on the PLC; be a separate hand-held device; or be a specially configured personal computer.

PID Controller (Proportional-Integral-Derivative Controller): This controller takes a measured value and determines the difference between it and a reference set point value. The difference is used to adjust the process to bring it closer to the desired set point. A PID controller can adjust processes based on proportion, integral, and derivative values. The proportion value determines the reaction to the current difference, the integral value determines the reaction to a sum of recent differences and the derivative value determines the reaction to the rate of change of difference.

Port: 1) Hardware Port = Connector used to access a system or circuit. Usually ports are used for the connection of peripheral equipment. 2) Software Port = Virtual data connection used by programs to exchange data without going through a file or other temporary storage location.

PPI (Point to Point Interface) Protocol: Used for communications between S7-200s in pure S7-200 networks via the integrated 9 socket D sub EIA-485 (RS-485) port.

Profibus (Process Field Bus): Popular type of field bus used in factory automation. It was developed in Germany and has more than 60% of the European automation market. It is included in the IEC 61158 and IEC 61784 standards.

Profibus DP (Process Field Bus Decentralized Periphery): Profibus used for fast input and output connections to sensors and actuators in factory automation.

Profinet: Industrial Ethernet standard for automation.

Protocol: Set of rules governing communication between electronic devices.

PT (Preset Time): Time when a timer turns on.

PTP or PtP (Point To Point): see PPI

PV (Preset Value): Value a counter counts to.

Rack: Framework or chassis that houses PLC modules. From a programming prospective, a single PLC framework or chassis sometimes contains more than one rack.

Relay: Electrically operated switch. Typically, when voltage is applied to a relay's coil, a magnetic field is produced to move an iron core that mechanically opens or closes electrical contacts. Typical voltage ratings for relay coils are between 12 and 115 volts, AC or DC. Usually the current rating of a relay's contacts are a few amperes or less.

In many circuit diagrams, circles represent relay coils and parallel lines represent relay contacts. A relay's coil is designated by the same letters as its contacts. Voltage applied to the A coil will close the A contacts.

There are also more complicated electromagnetic and solid-state relays. Some examples are time-delay relays, voltage level sensing relays, and current level sensing relays.

Register: Data storage location in a computer or PLC. This could store a word or group of words.

Retentive Register: Data storage location that retains its data during a power down.

RS-485: See EIA-485

RTD (Resistance Temperature Detector): Temperature sensor that predictably changes in resistance with temperature change. They usually use platinum and so are sometimes called PRTs (Platinum Resistance Thermometers).

Scan Time: Time to read all inputs, execute the control program, and update all input and output statuses. It is the time required to activate an output that is being controlled by a PLC. If the scan time is too long, a PLC may not be able to successfully control a process.

SFC (Sequential Function Chart): Graphical PLC programming language.

STL (Statement List): Also called IL, Instruction List. This is a text-based PLC programming language.

Symbolic address: Address that uses a combination of alphanumeric characters. Symbolic addresses can be global or local depending on how they are specified. Global addresses are common to all subroutines. Local addresses apply only to the subroutine in which they are used. Symbolic addresses are assigned to Absolute Addresses.

Upward Migration: Term that indicates that a PLC can do everything that its previous model could do, plus some additional instructions.

USS (Universal Simple Serial Interface): Siemens protocol for inverter control.

Word: Unit of binary data storage that is twice the size of a Byte. Modern computers have Word sizes of 16, 32, and 64 bits. With Step 7-Micro/WIN a Word has 16 Bits.

9.5 TECHNICAL SPECIFICATIONS COMMON TO TODAY'S S7-200S (CPUS 221, 222, 224, 224XP, & 226)

32-bit floating-point arithmetic in accordance with IEEE norm
Fully configurable, integrated PID controllers, up to eight independent controllers
Bit processing speed 0.22 µs
2 time-controlled interrupts (cycle between 1 and 255 ms at 1 ms resolution)
Maximum four hardware interrupt inputs (edge detection at inputs)
256 flags
256 timers
256 counters
4 to 6 high-speed counters (depending on CPU), maximum 30 kHz (or maximum 200 kHz with CPU 224 XP)
2 pulse outputs (pulse-width or frequency modulated), 20 kHz each for DC versions (100 kHz with CPU 224 XP)
Retentive (non-volatile) program and data memory
Retentive storage of dynamic data in the event of a power failure, non-volatile via internal high-performance capacitor and/or additional battery module, loading of data lock with STEP 7-Micro/WIN, TD 200C or by user program to integrated EEPROM
Typically 200 days buffering of the dynamic data with battery module
Integrated EIA-485 (RS-485) communications interface supporting the following operating modes: PPI master or slave/MPI slave/Freeport (freely configurable ASCII protocol)
Maximum baud rate: 187.5 kbaud (PPI/MPI) or 115.2 kbaud (Freeport)
STEP 7-Micro/WIN programming software supports standards such as STL, CSF, and LAD
Optional program memory module, programmable in CPU, for program transmission, data logging, recipe, & documentation
DC/DC/DC version uses: 24 VDC supply voltage, 24 VDC digital inputs, & 24 VDC digital outputs with 0.75 A capacity (parallel connection possible for higher switching capacity)
AC/DC/relay version uses: 85 to 264 VAC supply voltage, 24 VDC digital inputs, & 5 to 30 VDC or 5 to 250 VAC digital outputs (2 A capacity with the relay output version)

9.6 TECHNICAL SPECIFICATIONS SPECIFIC TO THE CPU 222

8 integrated digital inputs/ six integrated digital outputs
Maximum of 40 digital inputs with expansion modules
Maximum of 38 digital outputs with expansion modules
Maximum of eight analog inputs with expansion modules
Maximum of four analog outputs with expansion modules
4 KB program memory
2 KB data memory
50 hours storage of dynamic data via a high-performance capacitor
4 two-phase high-speed counters at 30 kHz or alternately two quadrature high-speed counters at 20 kHz
One EIA-485 (RS-485) communications interface
PPI master/slave protocol
MPI slave protocol
Freeport (freely configurable ASCII) protocol
Optional PROFIBUS DP Slave and/or AS-Interface Master/Ethernet/Internet Modem
One integrated 8-bit analog potentiometer (for commissioning, value change)
Optional real time clock
Integrated 24 VDC sensor supply voltage, maximum of 180 mA
Dimensions (W x H x D) 90 mm x 80 mm x 62 mm

9.7 EXPANSION MODULES AVAILABLE FOR CPUS 222, 224, 224XP, & 226

Digital input module, eight inputs, 24 VDC
Digital input module, eight inputs, 120/230 VAC
Digital input module, 16 inputs, 24 VDC
Digital output module, eight outputs, 24 VDC
Digital output module, eight outputs, relay
Digital output module, eight outputs, 120/230 VAC
Digital output module, four outputs, 24 VDC, 5 A
Digital output module, four outputs, relay, 10 A
Digital input/output module, four inputs/4 outputs, 24 VDC
Digital input/output module, four inputs 24 VDC/4 outputs relay
Digital input/output module, eight inputs/8 outputs, 24 VDC
Digital input/output module, eight inputs 24 VDC/8 outputs relay
Digital input/output module, 16 inputs/16 outputs, 24 VDC
Digital input/output module, 16 inputs 24 VDC/16 outputs relay
Digital input/output module, 32 inputs/32 outputs, 24 VDC
Digital input/output module, 32 inputs 24 VDC/32 outputs relay
Analog input module, four inputs, 12 bit
Analog input/output module, four inputs/1 output, 12 bit
Analog output module, two outputs, 12 bit
Analog input resistance temperature detector module, two inputs
Analog input thermocouple module, four inputs
Positioning expansion module for stepper motors or servo drives
SIWAREX MS Micro Scale weighing module
PROFIBUS DP communication module
AS-Interface master communication module
Modem EM communication module
Industrial Ethernet communications module
GPRS modem SINAUT MD communication module
Antenna ANT communication module

Printed in the United States
124750LV00002B/248/A